Yale Agrarian Studies Series

James C. Scott, series editor

The Agrarian Studies Series at Yale University Press seeks to publish outstanding and original interdisciplinary work on agriculture and rural society—for any period, in any location. Works of daring that question existing paradigms and fill abstract categories with the lived experience of rural people are especially encouraged.

James C. Scott, *Series Editor*

For a complete list of titles in the Yale Agrarian Studies Series, visit https://yalebooks.yale.edu/search-results/?series=yle10-yale-agrarian-studies-series.

LOVE FOR THE LAND

LESSONS FROM FARMERS WHO PERSIST IN PLACE

BROOKS LAMB

Yale

UNIVERSITY PRESS

New Haven and London

Published with assistance from the Yale School of the Environment,
and from the foundation established in memory of Calvin Chapin
of the Class of 1788, Yale College.

Yale University Press books may be purchased in quantity for educational,
business, or promotional use. For information, please e-mail sales.press@
yale.edu (U.S. office) or sales@yaleup.co.uk (U.K. office).

Set in Gotham and Adobe Garamond type by IDS Infotech Ltd.
Printed in the United States of America.

Library of Congress Control Number: 2022945750
ISBN 978-0-300-26744-0 (hardcover : alk. paper)
ISBN 978-0-300-28010-4 (paperback)

A catalogue record for this book is available from the British Library.

10 9 8 7 6 5 4 3 2 1

This book is dedicated to the people and places—farms, forests, friends, and family—who have taught me about imagination, affection, and fidelity.

I'm grateful for them all—especially Mama, Daddy, and Regan.

Throughout this work, I write about small and midsized farmers' stewardship. In some cases, these farmers and their families have been on the same land for well over a century. But it's important to recognize that long before these farmers and their ancestors came along, Indigenous peoples stewarded these places. Through a litany of injustices, they were forced from these lands. This history should not be forgotten.

Contents

Preface

Hal Manier owns a farm in Holts Corner, Tennessee. The Maniers have been tending this same piece of earth since 1919, when Hal's grandparents bought the land. At about ninety acres, the farm is not large. Although fertile enough to support a healthy herd of cattle, the soils are not rich. The barns and buildings are functional but weathered. With its small creek and rolling hills, the place is pretty but far from perfect. To Hal, it is home.

I have known Hal my whole life. Like his, my family has long been rooted in Holts Corner. Our small farm, which we have been caring for since 1892, is only a few miles from Hal's front door. Despite a long history of our families swapping farm work and attending church together, I didn't get to know Hal well until about five years ago. At that time, I was working for The Land Trust for Tennessee, helping farmers and landowners across the Volunteer State permanently protect their fields and forests from sprawling residential and commercial development. At a community gathering on a neighbor's farm in Holts Corner, Hal pulled me aside and asked some questions about conserving his land. He invited me to his place the next week, and we talked about the ins and outs of land conservation while sitting on the front porch.

After some reflection, Hal decided to donate a conservation easement on his farm, which meant that he would sever the development rights from the property so it would be open space forever. The process took a year of work: document drafting, report writing, boundary marking, and more. Toward the end of 2018, Hal and I met at the Marshall County Public Library, where he signed all the necessary documents. After the easement was recorded with the Register of Deeds, the Manier family land was officially conserved. Many farms have been paved over in the name of progress—especially in communities like Holts Corner, which are less than an hour from booming cities like Nashville—but Hal's land will not see that fate.

Hal was delighted after the easement was finalized. He took comfort in knowing that the farm can never be sacrificed for a subdivision or strip mall. Yet he was also aware that many people may not understand his commitment to conservation. They may even be baffled by it. Because the farm's future use was forever constrained by the easement, its potential sale price and value as collateral were drastically reduced. In the eyes of the wider world, it was devalued. Hal's loyalty to the land included significant financial sacrifice.

Hal is in his midseventies and has no children of his own. He has no nieces or nephews, and his one sibling is unable to tend the farm. He knew that the land may soon pass out of his family, that it could belong to someone other than a Manier for the first time in a century. So as we were wrapping up the land-protection process, Hal penned a letter to future owners of the farm, whoever they may be. He wanted people to understand why this place is so important to him. He wanted others to understand why he decided to protect it. With Hal's permission, I have included part of what he wrote:

> This land was always a refuge for me, a place to call home, and remains so to this day. Mark Twain said upon reaching 70 that he was approaching middle age. I suppose that applies to me as well. It occurred to me the other day when I was in a reflective mood that I am the only one on God's green earth who knows where each buried

rock that can hang a plow is located, where the wet spots are that should be left alone until dry weather, where the six wells are (some hidden now from disuse), where the best fishing hole in the creek to catch perch is, where all the family pets are buried, where the big weeping willow was in the yard, where the old cabin was located, where the old Jersey cow with the new-born calf "wallered" me in the ground when I was a six year old boy, and all those places and times my parents and grandparents taught me to love this land. I am the only one who knows these things, or cares.

My sister or I will be the last of my line to live on this place. . . . It is my fervent desire that those who own it after us will, as Ralph Waldo Emerson suggested, "let the land own them"; that they will let the land become a part of them and they a part of it.

Hal reveals a deep attunement to the family's farm, a localized knowledge earned and learned through countless experiences on, and "wallered into," the ground. Beyond this deep familiarity with place, he speaks to the attachment he has nurtured with the land. Physically, mentally, and spiritually, the farm is part of him.

Some people talk about love for the land superficially, yet Hal describes it in an enduring and specific sense. He explains that his family taught him to have affection for the farm. Clearly, their lessons stuck. Because of his intimate knowledge of, attachment to, and love for the farm, Hal has long acted as a steward. In an ultimate act of devotion to place, he ensured that the land will survive long after he's gone.

In many ways, Hal Manier is a rare and special character. He is, at least in my eyes, an exemplar of care, connection, and conservation. But he is also typical. He represents many other determined and devoted stewards who have nurtured relationships with place. Small and midsized farmers in rural communities across the United States and beyond are mirroring Hal's words and deeds.

In my two years working with The Land Trust for Tennessee, I met dozens of farmers who share Hal's commitment to the land. Around kitchen tables, on truck tailgates, and in feed stores, these farmers described their reasons for persisting in place. Most of the time, they weren't farming for money, or at least not much of it. Given our current agricultural system in the United States, their farms, tractors, bank accounts, and credit lines were far too small for that. Actually, many of them worked off-farm jobs in addition to raising crops or cattle so they could make annual farm payments, cover property taxes, and provide for family. These people were farming because of their attachment to place. They were farming because of love: for the land and the work, the community and the connection.

For many of the people I met, farming meant sacrifice. It meant, as Hal wrote elsewhere in his letter, "blood, sweat, tears, pain, and hard work—yes, hard work." But like Hal, their place was part of them, an integral element of their identity and purpose. While only a small percentage of these farmers decided to protect their land via conservation easements—many couldn't withstand the economic sacrifice of donating their development rights, and some didn't want to constrain their heirs—their commitments to the land were clear. From county to county, I began piecing together observations into an informal ethnography of sorts. A common thread of faithful care bound many folks from different places with different backgrounds, different ideas, and different political views together.

Beyond the moving words and stories I heard while working in a professional capacity, I have also witnessed the power of close-knit people-place relationships in my own life. Over the years, I have nurtured a deep relationship with my family's small, Tennessee farm. While hauling hay, feeding cattle, cutting tobacco, disking fields, picking vegetables, and sowing seeds, I came to know our place, and I learned love for the farm from my family. With that love comes a welcome obligation to care for the land, a joyful duty I can never shake. These seventy-five acres have shaped me. I am who I am because of this place and the people who call it home.

These experiences led me to explore enduring commitments to place in detail. To many people, these commitments might at first seem quaint or even nostalgic—at odds with "advanced" or "modern" methods of farming and an economy that sees "higher" and "better" uses for land. But the latest research only confirms what smaller farmers and rural communities have long known. Millions of acres are being paved over year after year, lost to farming forever, and much of what passes for farming today is slowly devastating the land and its capacity to sustain us. Numerous other harms follow from there. We are forgetting how we are fed. We are losing touch, quite literally, with the earth.

With all of this in mind, I wanted, even needed, to better understand why some farmers are willing to devote their money, their labor, their bodies, their time, and more for the sake of good and loving stewardship. This book—envisioned while traversing the backroads and pastures of rural Tennessee and penned during my time in graduate school at Yale School of the Environment—is the result of my efforts.

From the earliest pages, my research is anchored in the words and writings of Wendell Berry, who supplies the language needed to comprehend and communicate enduring commitments to place. Specifically, I examine how Berry's notions of imagination, affection, and fidelity—which I position as virtues, or enduring character traits that encourage people to think and act in certain ways—can motivate some small and midsized farmers of diverse backgrounds to sustain their stewardship, even and especially in the face of intense adversity. While agricultural adversity can take many forms, I pay particular attention to the systemic challenges presented by farmland loss from real estate development, rampant and relentless consolidation in the agricultural field, and—for Black farmers—historical and contemporary racism.

Through grounded research in two different locations in Tennessee, I reckon with this adversity through personal experiences. When conservationists talk about farmland loss in the United States, we often speak in statistics. Figures are important, but they cannot communicate all the impacts of farmland conversion. Rather than rely on numbers alone, I share emotions—heartbreak, frustration, and confusion; anger, despair, and defiance—from farmers and other community members who have witnessed the transformation of their once-rural communities. I convey troubling firsthand accounts of consolidation's impact on smaller, less wealthy farmers, showing how, as one farmer and community leader said, "the farm program we've got today put the little guy out of business." I also highlight the devastating impacts of systemic and individualized racism on Black farmers, supplementing individual experiences of injustice with historical context that speaks to centuries of oppression in American agriculture.

After discussing these challenges, I explore resilience. Farmers' own words and actions show how the virtues of imagination, affection, and fidelity have motivated steady stewardship. With regard to stewardship, I focus less on specific practices and more on presence. Sound farming practices are important, and other books, films, and articles discuss them in depth. Some of the people I interviewed employ such methods as rotational grazing, cover cropping, no-till planting, organic farming, and direct marketing, among other regenerative approaches like farm diversification, crop rotation, and input reduction. But in the face of great hardship, persistence and presence alone are powerful for many smaller-scale farmers. In a flawed system in which success can be equated to survival, perseverance and "hanging on" speak volumes to stewardship, which is why I've focused my efforts here.

Many of the farmers interviewed for this book express an unwavering commitment to tending their farms, even when it requires tremendous sacrifice. Several told me that they have turned down tens of thousands of dollars per acre—prices that would make these working-class people instant millionaires—in order to hang on to the land. They also mentioned refusing to enlarge their farms to unhealthy scales. In addition to resisting calls to sell or expand, Black farmers described fighting through loan denials, unjust property laws, and subtle and overt discrimination so they could stay in place. Against market forces, hostile policy, and contrary conventional wisdom, these farmers won't get big, and they won't go out. Imagination, affection, and fidelity help explain why.

The book closes by describing some strategies that local, state, and federal leaders can adopt to benefit rural communities and the earth itself. We must chip away at the structures and barriers that keep stewardship virtues from flourishing more widely by buttressing personal responsibility and place-based devotion with thoughtful policy. With care, courage, and hope, we can leverage love for the land and forge a better future for farming, one that sees the virtues of imagination, affection, and fidelity as assets that enrich and empower rather than liabilities that demand constant struggle. We can value places and people over the profits of a handful of wealthy farmers and massive corporations.

Love for the Land argues that we have much to learn from the farmer-exemplars who practice imagination, affection, and fidelity. Their commitment to and persistence in place is illuminating and inspiring. If we hope to address our most pressing agricultural, ecological, and cultural challenges—if we hope to forge stronger and more empathetic bonds with our neighbors, our places, and by extension our planet—we should pay attention to their instructive example.

Love for the Land

The Virtues of Imagination, Affection, and Fidelity: A Philosophical Foundation

In April 2012, Wendell Berry delivered the forty-first annual Jefferson Lecture in Washington, DC. Sponsored by the National Endowment for the Humanities, this lecture is recognized as "the most prestigious honor the federal government bestows for distinguished intellectual achievement in the humanities" and has been won by such figures as Robert Penn Warren, John Hope Franklin, Toni Morrison, Henry Louis Gates Jr., and Martha Nussbaum.[1] With this honor came a big stage and a chance to make an impact.

Recognizing the gravity of the situation, Berry spoke about stewardship, a worthy topic given the many environmental crises we face.[2] But he didn't discuss it in a typical way, refraining from calls for landscape-scale conservation, climate activism, or environmental legislation. And he didn't equate stewardship with calculated, impersonal "management" that aims above all else to extract consistent wealth from resources, as others have done. Instead, Berry concentrated on the aspect of good stewardship that he deems most important, one he has devoted his life to nurturing and communicating: affection's role as the "primary motive for good care and good use" of the earth. Anticipating pushback, Berry noted that "there is some risk in making affection the pivot of an argument. . . . The charge will be made that affection is an emotion, merely 'subjective.' . . . But the risk, I think, is only that affection is personal." Bucking the norms of mainstream environmentalism and conservation, Berry focused on intimacy rather than policy, on the personal rather than the public, on the local rather than the regional, national, or global.[3]

To make his case for affection, Berry presented the paradox of American agriculture's relation to the environment. On the one hand, he argued, there are farmers who are relentless exploiters of the earth, who take as much from their land as possible to increase their yields and profits. Eagerly following models and prescriptions issued by agribusiness experts and the government—prescriptions like former secretary of agriculture Earl Butz's command to farmers to "get big or get out" and recent secretary of agriculture Sonny Perdue's echo, "In America, the big get bigger and the small go out"—these farmers have little affection for place. Shaped by larger cultural, economic, and agricultural systems, they direct their love and energy mostly toward profit and competition, a dynamic Berry acknowledges in the opening lines of his poem "Manifesto: The Mad Farmer Liberation Front." Borrowing language from his former professor Wallace Stegner, Berry referred to these people as "boomers" who look to "pillage and run." These words are not hyperbolic. Evidence shows that many large-scale, overly intensive, industrial farmers and the structures that support them are indeed plundering communities, causing broad ecological, environmental, economic, social, and cultural harm. And while the scale and intensity of this harm has worsened in recent decades, these types of farms were wreaking havoc long before Earl Butz's mandate. Indeed, as rural writers like Grace Olmstead and Charles Thompson Jr. have shown through gripping memoirs, agricultural consolidation is leading to the dissolution, desertion, and destruction of many small, tight-knit, once-vibrant rural places.[4]

On the other hand are farmers whom Berry and Stegner call "stickers," or people who understand their land and commit to caring for and loving a place. To this group, Berry assigns agricultural icons like Aldo Leopold—who developed his idea of the land ethic only after he became the caretaker of a Wisconsin farm—Liberty Hyde Bailey, and Albert Howard. He also highlights other everyday farmers who nurture caring relationships with the land rather than use it exclusively for short-term financial gain. These farmers refuse to either get big *or* get out, thus rejecting the orders of Butz and Perdue and focusing instead on nurturing the health of their places and their communities. They are often motivated "by such love for a place and its life that they want to preserve it and remain in it."[5]

Berry's descriptions of these nurturing farmers' stewardship motivations are the focus of this chapter and the foundation of this book. Here, I argue that affection is far more than an emotion. It is best understood as a practical virtue, or an enduring component of a person's character that steers actions toward positive ends. A fleeting or sentimental feeling does not have the sustained force needed to encourage a person to be a "sticker." A virtue—which is a hard-earned habit of character—does. I also argue that imagination, which enables affection to flourish, is a virtue. So too is fidelity, the much-needed manifestation of affection. To make this case, I first share a working definition of virtue, using ideas from ancient and contemporary ethicists. Then I anchor imagination, affection, and fidelity as core virtues necessary for good stewardship.

After discussing these three stewardship virtues through a philosophical lens, I lift up a few on-the-ground examples of imagination, affection, and fidelity. I start by discussing how these virtues have motivated Berry himself to be a steward of his small Kentucky farm and then explore some of the essays Berry has written about visiting places cared for by other nurturing farmers. Berry is not a trained anthropologist or sociologist. Still, his limited ethnographic efforts begin to illustrate the power of imagination, affection, and fidelity.

With these examples in view, I suggest that a more in-depth study must be done to investigate the role that these virtues play in motivating some farmers to practice good stewardship. Berry's life and observations are useful, but we need a thorough, on-the-ground exploration of imagination, affection, and fidelity to determine the role these concepts play in cultivating and sustaining stewardship.

Virtues, Environmentalism, and Agriculture

Before we can understand the concepts of imagination, affection, and fidelity as virtues, we need to establish what virtues are. At their core, virtues are deep, enduring character dispositions. As Julia Annas states in her book *Intelligent Virtue*, "A virtue is a lasting feature of a person, a tendency for the person to be a certain way." If a person has the virtue of courage, this line of reasoning goes, they will not just be courageous on occasion or when it's convenient. Courage will be an essential part of their character, one that endures through time and space and meets the needs of various moments.[6]

In addition to helping define a person's identity, a virtue is also a trait of moral excellence. The renowned virtue ethicist Linda Zagzebski writes that this idea of excellence is one of the few elements of virtue theory that is not controversial among scholars. As a key component of moral excellence, virtues must be ordered or directed toward beneficial ends. These ends—or what some scholars call "goods"—should enhance a person's individual moral standing as well as make the world a better place. Virtues should serve a twofold practical purpose.[7]

We can take generosity—a widely recognized virtue—as an example. If a person is generous because they will be rewarded or recognized for their generosity or because some external force is demanding generosity, they are not acting in accord with virtue. Although they might do something nice for another person, the end in this case—

external reward or meeting a requirement—is not reflective of moral excellence, and the action's benefits are, even if useful, misguided. But if they are generous because they have a character disposition that values generosity for its own sake and frequently act to bring about compassion and relieve suffering for others, then they have served both themselves and the wider world by practicing the virtue of generosity.[8]

Perhaps the most important element of virtue is that it must be acquired. Virtues do not arise naturally. As Zagzebski notes, "A human person's moral identity is intrinsically connected with a series of experiences of interaction with the world around her. . . . The virtues of human beings in the ordinary human environment are acquired at least in part through the acquisition of certain habits." Aristotle himself posited that virtue must be acquired and practiced. In book 2 of his *Nicomachean Ethics*, the philosopher wrote that virtue is the result of teaching and habit. "Neither by nature . . . nor contrary to nature are the virtues present; they are instead present in us who are of such a nature as to receive them [i.e., being taught about virtue], and who complete or perfect themselves through habit [i.e., repeatedly practicing being virtuous]."[9] In other words, to be virtuous requires a willingness to learn from others and a desire to practice. For these reasons, cultivating virtue takes time, attention, and patience.

While virtue must be acquired by practice, it's wrong to assume that these habits become static or routine. By "habit," neither Aristotle nor other virtue ethicists are referring to something that we do without thinking, such as tapping our toes during a song, locking our car doors in a parking lot, or brushing our teeth before bed. Instead, we become habituated to virtue by reasoning our way through various situations and trying to act in alignment with the virtue in question.[10] The act of *trying* to be virtuous is the habit that must be formed.

Some of the situations we encounter will be simple, making it easy to discern how to act excellently. Other situations will be more challenging, which may cause us to struggle. These struggles should not be seen as weakness or ethical inferiority. Rather, the task of struggling to act virtuously is central to becoming a morally excellent person. "The idea of somebody acquiring moral character without struggle," Robert Roberts argues, "seems not only psychologically, but logically amiss." In a lecture especially relevant for the discussion of struggle, virtue, and agriculture, the first-century Stoic philosopher Musonius Rufus states that the best job for a philosopher is farming, partly because it requires one to endure repeated hardship and practice perseverance in both body and mind.[11] Farmers have ample chances to practice virtues.

An important component of the struggle to achieve moral excellence for farmers and nonfarmers alike is that we will sometimes fail. No human can be virtuous at all times. A courageous person may occasionally act cowardly. A generous person may sometimes be selfish. Yet as long as we continue to habituate ourselves toward good and virtuous ends, we will continue to move toward excellence, bettering ourselves and our communities in the process.

In short, a virtue must be a deep and lasting feature of a person's character, one that is acquired by habit and, when possible, learned from others. It must also be directed toward good ends. Further, when a virtue guides our actions, these actions should serve both ourselves and others. While "the word virtue has an archaic ring to some ears and has dropped out of use entirely in many quarters," writes Zagzebski, this concept has immense practical importance when understood as just described.[12]

Some of the United States' most well-known environmentalists— people like Henry David Thoreau, Aldo Leopold, and Rachel Carson— have recognized the utility of employing virtue to guide ecological care. And while recent discussions in environmental ethics have centered on rights-based approaches to ensure environmental protection or focused on legislative theory and action to tackle ecological problems, a cohort of scholars and writers remains committed to advocating for the development of personal eco-virtues. Ronald Sandler, for example, argues that virtues "can be the basis for an inclusive environmental ethics that accommodates the richness and complexity of human relationships and interactions with the natural environment and provides guidance on concrete environmental issues." Similarly, Louke van Wensveen and Paul Thompson show that individual character has an important role in addressing urgent environmental and agricultural problems. In a thorough edited volume, Heesoon Bai, David Chang, and Charles Scott argue that ecological virtues are the key to living well in our new era, while Jason Kawall marshals similar claims in *The Virtues of Sustainability*. All these scholars—not to mention countless Indigenous thinkers, who have long avowed similar ethical characteristics—rehabilitate virtues as active and useful instead of dusty and abstract.[13]

Wendell Berry also belongs to this group of environmental writers who embrace virtues as valuable on their own and as paths toward good stewardship, even if he prefers not to use academic language to do so. As Kimberly Smith notes in *Wendell Berry and the Agrarian Tradition*, "the academic community . . . is not Berry's primary audience. He is writing to a broader public, to everyone who is or should be concerned with achieving a healthy and sustainable relationship to the living world." While few scholars have explicitly identified his concepts of imagination, affection, and fidelity as a network of stewardship virtues, other elements of his thought have been positioned in this way. For instance, van Wensveen labels Berry's understanding of discipline as a core environmental and agricultural virtue since it helps a farmer focus on the details needed to nurture a healthy farm, prioritize conservation, and exercise productive self-criticism. Motivated by internal disposition, the disciplined farmer aims for personal and communal excellence. Thompson makes comparable claims about Berry's writings on industriousness and positions it as an essential agrarian virtue.[14]

Similarly, Michael Lamb argues that Berry's "difficult hope" should be understood as a virtue. Through a close reading of Berry's essays and a personal interview, Lamb shows that hope, as Berry defines it, is not a fleeting emotion and should not be equated to "wishful thinking." Rather, when properly practiced, hope is an enduring trait directed toward doing good work. Hope—which Lamb describes as difficult but necessary to achieve, falling in line with Roberts's previously discussed struggle theory—is especially important for farmers to nurture given the tough tasks they face daily. Without it, few farmers would ever see a harvest. They might struggle to find the will to plant a crop in the first place. Hope is key for environmentalists, too. Facing such pressing issues as species extinctions and mass deforestation, ocean pollution and climate change, environmentalists must maintain hope to continue their good and necessary work.[15]

These portrayals of Berry's notions of discipline, industriousness, and hope as practical virtues are helpful for understanding how imagination, affection, and fidelity will be similarly positioned. By situating stewardship virtues in this way, I will show how and why these concepts encourage some farmers to be nurturing caretakers instead of exploiters.

Berry's Virtues of Imagination, Affection, and Fidelity
Imagination

Imagination is the catalyst that enables affection and fidelity to flourish. Without it, the other two virtues wouldn't materialize. However, imagination is also an important stewardship virtue on its own, independent of its role in stoking affection for and fidelity to place. But why is imagination—a term often associated with a child's ability to dream up fantasies of dancing unicorns, talking dogs, or magical forests—important for promoting agrarian and ecological stewardship?

In "It All Turns on Affection," Berry says that imagination, as he understands it, is unlike the fantastical definition of the word often proclaimed in popular culture. "To take it seriously," he writes, "we must give up at once any notion that imagination is disconnected from reality or truth or knowledge. It has nothing to do with clever imitation of appearances or 'dreaming up.' " Instead, "imagination thrives on contact, on tangible connection" with a place.[16] It must exist in real life.

For a farmer to acquire this authentic imagination of a place—to understand its soils and topography, its climate and character, its ecosystems and people, its past and potential—they must work toward knowing it well. Deep awareness of the farm must become a part of their very self, a core aspect of their identity. Even from thousands of miles away, they should be able to shut their eyes and picture their farm, summoning all the richness in detail that makes their place unique.[17]

An understanding of place this intimate requires time and commitment. It may arise through work on the land: dozens of drives through a field or a forest, hundreds of walks to feed livestock or mend fences, years—sometimes decades—of energy and effort spent tending a specific place on earth. It might also be engendered through lessons from other farmers or family members who have already cultivated a sense of imagination for place. Learning about the land and how to tend it from others and spending ample time on it oneself empowers a person to truly "see" a place. Imagination, writes Ragan Sutterfield, a minister and writer who studies Berry's work, enables a person to understand "how things fit together in their past and present, the relationships of the present, and the future trajectory that reality might take."[18]

If Berry's notion of imagination is still difficult to understand, especially given the word's popular meaning, it's helpful to consider different terms that could be used as partial synonyms. "Familiarity," a word Berry at times uses to communicate a close knowledge of place, is one potential alternative, though it lacks the force and richness of "imagination." "Attention," which Jeffrey Bilbro discusses extensively in his book on virtues in Berry's literature, is also a comparable word. "Attunement" is another. The eco-feminist scholar Elizabeth Dodson Gray writes that attunement is much like active listening and should be considered one of the core environmental virtues. We must listen to the landscapes we are in, she argues, if we are to know, understand, and then treat them well. The similarities between Dodson Gray's "listening" and Berry's "seeing," while obvious, are worth noting explicitly. In the same way that Berry urges people to "see" and not just "look," Dodson Gray asks her readers to "listen" and not just "hear." Imagination and attunement align with the deeper charges of seeing and listening.[19]

Imagination—or attunement, attention, or familiarity—is knowing a place in the most intimate way possible and, from that knowledge, understanding what it needs and how to care for it. Nurturing imagination of a place in this way helps a farmer trend toward being a "sticker" because they are ever aware of the farm's condition. At its best, imagination yields wisdom that is practical, local, and applied. This evolving, place-based knowledge—which is a cornerstone of current conversations around "regenerative agriculture"—is different from the kind often generated by hard science. As Berry notes in "Renewing Husbandry," scientific knowledge alone does not often consider the intricacies of each individual farm or farmer. While it can be immensely useful when applied responsibly and not viewed as the end-all-be-all answer to every single situation, scientific knowledge is frequently generalized. The knowledge that comes from imagination, on the other hand, is specific and local. It arises from familiarity with a place and not from a far-away laboratory or textbook. This intimate knowledge enables a person to better understand how a place can flourish. It puts a person on the path to becoming a sticker, a nurturer, a person who, as Berry writes in *The Unsettling of America*, values health—their land's health, their own, their family's, their community's—more than profit and status.[20]

The kind of understanding that imagination is directed toward is like the localized knowledge of "metis" that James Scott describes in *Seeing Like a State: How Certain Schemes to Improve the Human Condition Have Failed*. Scott states that metis is understood as "acquired intelligence in responding to a constantly changing natural and human environment." It "depends on an exceptionally close and astute observation of the environment," a knowledge born of attentiveness and awareness that sees farming as an art rather than a general, scientific industry. He goes on to explain that fostering metis requires time and practice. This knowledge can't be developed overnight. Scott's focused commentary on the metis needed for farming is particularly relevant and insightful: "All farming takes place in a unique space (fields, soils, crops) and at a unique time (weather pattern, season, cycle in pest populations) and for unique ends (this family with its needs and tastes). A mechanical application of generic rules that ignores these particularities is an invitation to practical failure, social disillusionment, or most likely both." Like the deep knowledge that Berry's imagination is directed toward, Scott's explanation of metis serves the practical purpose of helping a farmer provide good care for the land, for themselves, and thus for others.[21]

Given this understanding, imagination can be positioned as a stewardship virtue in the following ways. First, it's an enduring element of a person's character, one that can stand the test of time. If a person has the virtue of imagination, they are deeply attentive to their place: to its history, its present condition, and its envisioned future. Further, imagination is ordered toward the excellent end of deep and usable knowledge. Acting on this knowledge not only makes a person a better farmer but also promotes agricultural, ecological, and economic stewardship that serves others and the well-being of the farm itself. After all, Berry says that "kindly use depends upon intimate knowledge, the most sensitive responsiveness and responsibility."[22] Finally, genuine imagination of a place is not something that arises in a person innately. It must be acquired by learning about the intricacies of a place from a more experienced person and becoming attuned to the land through sustained effort. Nurturing imagination requires discipline and work. Honing it must become a never-ending habit.

But how does imagination enable affection? In *The Place of Imagination: Wendell Berry and the Poetics of Community, Affection, and Identity*, Joseph Wiebe argues that true imagination of a place "evokes a response that forms a connection, a relationship, involving obligations that are felt, which is to say, constantly thought. . . . Imagination makes real the life of a place, which simultaneously reveals the various claims a particular environment makes on those who share its life."[23] When the life of a place is made real—when we come to know a place in the most intimate fashion and cultivate a relationship with it anchored in attunement and understanding—we often begin to have empathy and affection for the place. We're better able to love what we know.

A human analogy illustrates Berry's theory of imagination yielding affection. Affection for a place depends on imagination much like our love for a friend. Depending on our specific social, cultural, and geographic situations, we may come across dozens, hundreds, or thousands of people each day. Usually, we encounter these people in an impersonal way, passing them on a sidewalk or a highway, in a hallway or an airport. While we care about their well-being—we certainly don't wish them harm—we are not invested in them. We don't *know* them. Our friends, on the other hand, are people with whom we've spent an abundant amount of time. We know them on deep and personal levels. When something is bothering one of our friends, we can tell. When they are excited, we can sense it. We are attuned to their actions, their expressions, even their underlying mood.

Knowing a person in this way leads to affection for them. Of course, the process of developing affection for a friend, like for a place, is iterative. It's not like we reach a certain threshold of imagination and then qualify to love our friends. Affection builds as we nurture a relationship. But it is through the act of cultivating imagination that we first form deep and lasting bonds to others, bonds anchored in understanding. The same can be said for a farmer's imagination for their land. Without imagination, affection cannot arise.

Affection

Like Berry's understanding of imagination, his concept of affection is sometimes misunderstood. Whereas many people think of imagination as fantasy, affection is often perceived as a fleeting emotion rather than a persisting character disposition. As Norman Wirzba affirms, "it is easy to confuse [Berry's] affection with romantic bliss and ease." Berry himself recognizes that affection is generally viewed as a sentimental and simple feeling, which makes him hesitant to position it as the linchpin of stewardship. He knows that if he makes an argument based on emotion, his idea may be dismissed. Others, like bell hooks, have also been careful to position love as more than an emotion. In fact, hooks wrote an entire book on love, in which she argues that it requires great effort, attention, and practice.[24]

It should be noted here that emotion is not without merit. The opposite is true. Emotions inform and govern many aspects of our lives. They are part of what makes us human. In fact, research shows that we even act on emotions when deciding to care for the earth. In a case study that explored volunteers' motivations for working with an organization that focuses on native plant restoration along riparian buffers, researchers found that many people donated their time to the cause because their emotions compelled them to do so. Emotions that clustered around themes of responsibility, guilt, and pleasure were of particular importance. They even found that "love of nature," which sounds like affection, was a significant emotional motivator for getting involved.[25] Emotions can thus be significant for stirring actions that promote environmental care.

There is, however, a crucial difference between a motivator that encourages someone to volunteer on occasion with an environmental nonprofit and a motivator that sustains long-term, loving care for a place amid difficulty. While emotions may bring about bursts of short-term action, they are temporary. An investigation by Phillipe Verduyn and Saskia Lavrijsen found that most emotions last "anywhere from a couple of seconds up to several hours."[26] Given the fleeting nature of emotion, then, it can't be considered an effective means for inspiring enduring stewardship. For this end, a person needs something more.

Berry seems to understand the limits of emotion, which is why he is careful to paint affection as something beyond a simple feeling. Rather than an emotion that lasts for a few moments, affection "involves us entirely" across space and time. It becomes a component of a person's identity—a proactive element of character—and not a response to a situation. As Wiebe writes of affection, it is a "force, one that emerges from imagination, encounters various conflicts, and forms robust communal bonds" with a place.[27]

Importantly, affection endures through hardships. Berry cites a common feeling among farmers, especially those who engage in difficult physical work. At times, a farmer may be baking in the sun, scrambling to get their hay crop into the barn before the rains come. They may be feeding that same hay to their cattle six months later in the freezing cold, their hands and face numb from the howling wind and snow.[28] In these trying times, it is reasonable to assume that the farmer is experiencing a variety of emotions: dread, frustration, self-pity, annoyance, maybe even anger. No one would blame them for not feeling joy. But their guiding disposition—the reason they do this work in the first place, even if it may at times be unpleasant—is affection, nurtured through imagination of their farm and its needs and directed toward thoughtful care. Their work is governed by this active, enduring trait and not by reactive, fleeting emotions.

Berry addresses this idea of affection's role in promoting perseverance and care in the face of hardship in his essay "Conservationist and Agrarian." He asks, "Why do farmers farm, given their economic adversities on top of the many frustrations and difficulties normal to farming?" These difficulties are those of the sort described earlier: exhausting physical labor, long hours, brutal weather, never-ending work, uncertain and often-dim economic prospects, the ever-present possibility of failure. Immediately answering this question, Berry responds that "always, the answer is love. They must do it for love. Farmers farm for the love of farming. They love to watch and nurture the growth of plants. They love to live in the presence of animals. They love to work outdoors . . . maybe even when it is making them miserable."[29] Affection, or love, is the sustaining motivation that nurturing farmers use to advance the overall health of their land, their families, their communities, and themselves.

Cultivating affection that can endure through difficulties, however, takes work. As with imagination, a farmer must learn to love a place, both from their experiences on it and, if possible, from others who have loved it too. "Affection," Wirzba notes, "grows over time and is the effect of sustained commitment and involvement." It is partly for this reason that Berry advocates for the small and midsized family farm. In this setting, children are taught to imagine and have affection for the land by their guardians. They also have the opportunity to learn these virtues through working on the farm, either alone or with family. It is not necessary for a person to grow up on a farm in order to cultivate affection and become a good and nurturing steward.[30] Nor is it guaranteed that a child who grows up on a farm will develop affection for it. In fact—and for various reasons, as will be explored later in the book—many children who grow up on farms leave their family's land and farming altogether. But to become a devoted caretaker, it is paramount for a person to acquire love for a place in a deep and lasting sense, binding themselves to it through hard-earned affection.

With this understanding of affection in view, it is clear how the concept can be positioned as a virtue. Affection is a fully involved element of a person's character. Unlike an emotion, it's not merely a quick reaction to a given circumstance. It is instead an enduring disposition to be and act a certain way, and it is ordered toward the end of nurturing the health of a place and its people. Further, the virtue of affection must be acquired through habit. For farmers, it is important that these habits withstand and persevere through struggle, which they will inevitably encounter. Virtue may not always prevail. A bitter winter wind, the blazing summer sun, or brutal economic conditions may tempt a farmer to temporarily abandon or redirect their affection, to act in a way that is counter to it. But when the trait is deeply entrenched—if affection is indeed a core part of their identity—the farmer will continue working toward it in the future.

Although affection is of the utmost importance for its own sake, it—like imagination—is also a stimulant to another virtue. Affection engendered through imagination leads to fidelity. When we have nurtured an abiding love for a place, we feel compelled to care for it. Land becomes more than just a passive setting in which our lives unfold. Instead, it is a central character in the story, one that we cherish. Affection enables us to forge "a neighborly, kind, and conserving" relationship with a place, one grounded in understanding and empathy. Further, "without this informed, practical, and *practiced* affection," the prospects of a nurturing relationship with a place are nonexistent.[31] A relationship anchored in affection gives rise to a pleasing obligation of thoughtful, enduring care. In other words, it gives rise to fidelity.

To carry forward the same example used to highlight how imagination leads to affection, we can look at friendship to understand how affection stimulates fidelity. Friends are, as earlier explained, people whom we know and love. And because we love them, we hope and work for their well-being. We help them, when possible, to thrive and reach happiness and fulfillment, and they do the same for us. Our friendships, like virtues, are not perfect. We may argue with a friend. We may get angry with them, sometimes severely so. But if the friendship is anchored in understanding and affection, it can persist, even when confronted by difficulty or disagreement. It can continue to help both people flourish.

Fidelity

Unlike for imagination and affection, it is not uncommon for people to think of fidelity as a virtue. Often described as the disposition that governs faithfulness, loyalty, and lasting support, fidelity has characteristics that lend to its interpretation as a virtue, especially in light of the way that virtue is defined in this chapter. What *is* uncommon, however, is that fidelity be positioned as an ecological and agricultural virtue, a *stewardship* virtue. Generally, fidelity is practiced toward a person. It is known as the concept that determines how one person should treat another, particularly another person they love. It is not commonly directed toward our relationship with the earth.

Berry himself recognizes that fidelity is an important virtue for human relationships, and it's worth examining his conception of human-focused fidelity before moving on to the place-based iteration. Whereas I have used friendship as an analogy to communicate understandings of imagination and affection, Berry most notably sees fidelity's importance in the realm of marriage. "If fidelity is a virtue"—and he believes it is—"it is a virtue with a purpose." It must be radically practical. When resting on the foundation of imagination (knowing a spouse extremely well) and affection (loving that spouse deeply, so much so that the love becomes a central part of a person's identity), fidelity's purpose is to dispose someone toward loving care and long-term commitment. Fidelity is devotion that should nurture the well-being of both individuals in a marriage. It connotes a compact, a partnership, and a commitment wrapped into one, and it can "give us the highest joy we can know: that of union, communion, atonement (in the root sense of at-one-ment)."[32]

Given Berry's understanding of fidelity as a virtue in healthy marriages and its striking relevance to people-place interactions, it's not surprising that he uses marriage as a metaphor to describe the relationship between a good farmer and their land. Because fidelity—as rooted in imagination and affection—is at the core of both marriage and virtuous stewardship, "the analogy between marriage making and farm making, marriage and farm keeping, is nearly exact." Marriage and good farming bring their own versions of joy and pain. Both present hardships and happiness, and both require faithfulness and demand the same kind of responsibility. Whether caring for a person or a place, devotion is needed. "It is impossible to care for each other," Berry writes, "differently than we care for the earth."[33] Marriage, then, is a helpful analogy to understand fidelity's role in agricultural and ecological stewardship.

Analogies aside, fidelity—as engendered by authentic affection—is a stewardship virtue because it is, by nature, a deep and enduring disposition. It is directed toward the end of providing good care, of giving oneself to a place for the sake of stewardship, of advancing communal health. And it undoubtedly must be acquired and refined through habit. A farmer cannot become a faithful, devoted, and nurturing steward just by doing a few good acts. They must continually direct themselves toward loving care.

Of course, as with any virtue, a good steward will at times struggle with practicing fidelity. But here is where the importance of affection as a stimulant lies. If fidelity depended solely on the mind, it would become "perverted beyond redemption by understanding it as a grim, literal duty enforced only by willpower." When it stems from love, however, fidelity becomes a welcome responsibility. Rootedness in affection helps fidelity endure through the challenges that all caretakers face. Only with authentic affection stimulated through imagination can we bring about genuine hope for good and lasting stewardship, for it is love that enforces care.[34]

Virtues on the Ground:
Berry and His Amateur Ethnographies

The philosophical positioning of imagination, affection, and fidelity is important, and it serves as the basis of this book. But more important is how stewardship virtues operate in real life. If they "work" only in print, they serve no real function. In this section, I explore how these three virtues lead to good care on the ground by lifting up agrarian stewardship exemplars, including Berry himself.

Berry as an Exemplar of Stewardship Virtues

Wendell Berry grew up in rural Henry County, Kentucky—approximately fifty miles northeast of Louisville—and spent much of his childhood on his family's farm. "I was born to people who knew this place intimately," he writes, "and I grew up knowing it intimately." As he played and worked on the farm with siblings, parents, and grandparents, he continued to refine and deepen his imagination for the place. It became so clear that, when away from the farm, he could summon visions of fields and woods. Revealing forceful familiarity, he says that he could even recall the "casual locations of certain small rocks": "I had come to be aware of [the place] as one is aware of one's own body; it was present to me whether I thought of it or not." Berry learned, as he said in his Jefferson Lecture, to "see the place with a force of vision and even with visionary force."[35] With this strong imagination came affection for his place—his *home*—and it was this affection that bound him so strongly to the land.

Despite the love he had for this place, Berry left Henry County to attend the University of Kentucky in Lexington. From there, he went to Stanford to study with the revered writer and environmentalist Wallace Stegner. After finishing his studies in California and writing his first novel, Berry spent time in Europe on a Guggenheim Fellowship and then moved to New York City. Teaching creative writing at New York University, he had climbed to the brink of excellence. "I had reached the greatest city in the nation," he writes. "I had a good job; I was meeting other writers and talking with them and learning from them; I had reason to hope that I might take a still larger part in the literary life of that place."[36] From the outside, the stars seemed to be aligning for Berry the author. But Berry the person knew better. In what some people, including his fellow faculty members at NYU, saw as career suicide, Berry left New York to return to Henry County because he longed for home. Once there, he and his wife, Tanya, bought a small farm near his homeplace. Berry started teaching at the University of Kentucky, writing in a small shack, and caring for the land as a "part-time" farmer.

In an act of fidelity, Berry decided against conventional measures of success to return to his place and care for it. And while he felt that this was, to an extent, his fate, his motivation to return was rooted in affection. "I still had a deep love for the place I had been born in," he explains, "and I liked the idea of going back to be part of it again." The imagination and affection he had nurtured as a child led him home to be a caretaker of a small farm along the Kentucky River. Crucially, he continued to hone these virtues once arriving home, recognizing that if his imagination and affection were to be useful, they could not become stagnant.[37] His habituation of these virtues had to continue.

Berry has maintained fidelity to the place, caring for the land even as he ages. In his late eighties, Berry now also leans on others in his family and community to help him steward the farm. As he continues to practice imagination, affection, and fidelity, his life has become inseparable from the land. "I cannot identify myself to myself apart from [the farm]," he said almost fifty years after first returning to Henry County. "I am fairly literally flesh of its flesh. It is present in me, and to me, wherever I go."[38] As in a strong marriage, Berry and his place have become one, bound together in loving care.

Berry's Amateur Ethnographies

In addition to practicing these virtues himself, Berry has observed and communicated how other "stickers" have lived on and cared for their farms. These writings can be considered amateur ethnographies. The term "amateur" does not imply disrespect. As Berry himself writes, an amateur is simply someone who is "excluded from the professional 'field,' " someone who pursues a task not for money or fame but for love.[39] He shares stories from other farmers who care for their land and communities because it's something he *wants* to do, even though it's beyond his training and professional qualifications. And although studying stewardship virtues is not his goal in these pieces— he appears more interested in understanding the overall dynamics of the farms and people he visits—his work starts to show the real-world power of imagination, affection, and fidelity. Three of his farm forays, in particular, are useful for witnessing stewardship virtues in action.

First, we can examine Berry's experience of visiting an Ohio strip-mine-turned-farm in his essay "A Rescued Farm." This example is especially pertinent to a discussion of good stewardship because, ironically, Berry writes elsewhere that he "conceives a strip-miner to be a model exploiter."[40] The fact that a former mine is transformed into a small farm can be seen as a stroke of justice.

Berry visited this farm because the owner, Wally Aiken, invited him. In a letter to Berry, Aiken explained that he had already spent several years working on the place, doing his best to nurse it back to health. Once Berry arrives at the farm, he tours it with Aiken, who shows him the reclaimed land in all its detail. It is here that Berry first begins to describe Aiken's imagination of the land. Because Aiken immediately had to begin work to rehabilitate the land after purchasing it, he had to "come to know it in the process of changing it."[41] As a result, his imagination constantly evolved, even more so than it would on a typical landscape.

Because Aiken was committed to understanding his land, he began to see success. For example, after working on and walking the land, he devised a localized way to control soil erosion. Once he had leveled and graded a damaged plot, he began sowing a mixture of grasses and legumes on the bare dirt. Immediately after, Aiken spread onto the seeded ground "several hundred bales of spoiled and low-quality hay" that he had bought at a bargain from a nearby farmer.[42] Even after heavy rains, the rotten hay helped hold the seed in place, and soon, healthy roots kept the soil from washing away. Before too long, Aiken had a self-sustaining hillside. He had only been farming for a few years, and he had no formal training. But Aiken's knowledge of the land, developed and deepened through personal experience and related to the idea of "traditional environmental knowledge" put forth by some scholars, helped him understand how to care for it.

While Aiken was getting to know his land, he began to love it, even with its flaws and its required effort. Despite the long hours of picking up rocks, planting seeds, spreading spoiled hay, and more, he felt a certain satisfaction from being on the place. Aiken himself says this too: "It is nice to have something to devote oneself to," he told Berry, "to care about and be a part of." Aiken's love for the land and his relationship with it motivated his fidelity. At the end of the essay, pondering why Aiken acts as a careful steward of this rescued farm, Berry writes that "what [Aiken] has done is not 'practical' or 'economical,' as things are now reckoned, and certainly not easy. He has done it out of devotion to a possibility once almost destroyed in his place, and now almost recovered."[43] Imagination, affection, and fidelity were at play in the story of this mine-turned-farm.

Next, we should examine Berry's trip to a sheep farm in his native Kentucky, which he describes in "A Talent for Necessity." Henry Besuden of Clark County inherited a farm that was poorly cared for. His grandfather, who had apparently practiced none of the three stewardship virtues, had "corned [the land] to death," renting the land to big row-crop farmers who—following advice from extension agents and agricultural experts and aligning with conventional agroeconomic principles—planted fence row to fence row and exploited the land for profit as best they could. Indifferent, Besuden's grandfather did not even bother to "get up and see where [the corn] was planted." By the time Besuden became the owner and caretaker of the land, the farm was "covered with gullies, some of which were deep enough to hide a standing man."[44]

Besuden set about trying to care for the farm through cultivating his own understanding of the place and how it should be treated. Contrary to what the "boomers" did, Besuden reduced his plowed acreage, started using cover crops—a rare practice in the middle of the twentieth century and still too rare today—and even planted locust trees, which are viewed with disgust on many farms, to help reduce soil erosion and restore health through nitrogen fixation. He also brought sheep onto the farm. When carefully tended, these animals were easy on the land and helped rebuild its fertility. Besuden's time spent on the place, along with advice he had received from nearby kindly-use farmers, indicated that these practices could be helpful. In another example, Besuden made sure to see his pastures early each morning. On these visits, he was looking for the barely perceptible fresh, tender grass that would sprout through the old. Once he saw these new blades in a field, he knew it was time to bring his sheep there so that they could begin grazing the new growth and other fields could recover. This acute attention to detail made for healthier sheep, better nutrient cycling, and less-stressed land, as well as improved economic performance.[45] Besuden's imagination for the place and his understanding of its unique set of challenges helped him become a good caretaker.

Just as Aiken did, Besuden fostered affection for his land, and it was this love for the place that sustained his devotion. In a column for a sheep-farming magazine, Besuden wrote about the personal character and conviction needed for conserving soils: "This thing of soil conservation involves more than laying out a few terraces and diversion ditches and sowing to grass and legumes, it also involves the heart of the [person] managing the land. If he *loves* his soil, he will save it." Later, Berry writes that Besuden once "thought of numbering his fields, but decided against it—'That didn't seem fair to them'—for each has its own character and potential."[46] Besuden's quote about the love-driven motivation for conservation, along with his refusal to assign impersonal numbers to his fields, signals a fidelity to his place sustained by connection and affection. Over time, he cultivated a relationship with his farm, and he embraced the obligation to treat it well. His attunement and dedication made him a better farmer and his farm a better place.

Finally, we can explore the presence of imagination, affection, and fidelity in "Seven Amish Farms." Berry makes clear in this essay and elsewhere that he reveres the Amish and their way of life, and he writes about their farming frequently. "I do not recommend, of course, that all farmers should become Amish, nor do I want to suggest that the Amish are perfect people or that their way of life is perfect," he notes.[47] But he admires their culture, not least because it values the stewardship virtues.

In describing the differences between the northeastern Indiana Amish farms he visits, which are owned and operated by members of the Yoder family, and large industrial farms, Berry writes that it is like comparing "a desert and an oasis." While the industrial farms Berry speaks of are thousands of acres in size, the largest Yoder farm is only ninety-five acres. The farms' small acreages make them knowable, even more so when considering that the Yoders use horses instead of tractors to power most of their equipment, putting them quite literally closer to the ground. It's not practical for most current farmers to use these methods, but it works for this family. Because many of them have been walking behind a plow or planter since they were children, they have secured a careful and comprehensive knowledge of their land through hands-on experience.[48]

The Yoders have also been fortunate to have devoted teachers. Older farmers and family members have taught imagination to the next generation. Bill Yoder, the oldest member of the family at the time of Berry's visits, notes the understanding of land stewardship that his late father passed down to him, even quoting him verbatim to Berry in conversation. Bill has tried to return the favor, passing down localized wisdom to his own children. With help from exemplars and their own practice, the Yoders developed a strong sense of imagination.

The affection that the Yoders have cultivated, while stemming from imagination, also has a spiritual element, Berry explains. This family views the land as a neighbor, as an essential element of their community that is "to be loved as oneself." They "subordinate economic value to the values of religion and community," with which affection aligns.[49] In the latter quote, the Yoders' status as "stickers" is evident. Their fidelity to place—demonstrated by generations of family members who have lovingly cared for farms in their community, forsaking financial gains and easier livelihoods to embrace the call of stewardship—is anchored in understanding, respect, and love.

Toward an Ethnography of Imagination, Affection, and Fidelity

Berry's personal practice of stewardship virtues is instructive. His ethnographic forays also help communicate how these virtues arise on the ground. But neither Berry's example nor his amateur ethnographies can offer definitive evidence that imagination, affection, and fidelity are keys to good stewardship.

Why? Berry's life alone cannot be used to validate these virtues. If we want to test the idea that virtue-driven stewardship is widely applicable, we must look elsewhere to examine these theories in action. And while his amateur ethnographies illustrate the roles that imagination, affection, and fidelity play in motivating stewardship in select situations, they lack the organized analysis that a more formal and focused study can provide.

For these reasons, we need an ethnography—or a thorough and focused exploration of local people, places, and cultures—that studies farmers who nurture the land, who practice good stewardship and care while persisting through challenges and resisting top-down directives to "get big or get out." Connecting with and learning from farmers in their own communities via interviews and other qualitative methods would lead to a better understanding of their motivations and help determine if stewardship virtues are prevalent. Even asking farmers such seemingly simple questions as "Why do you continue to farm?" "How do you feel about your land?" and "Where does your agricultural knowledge come from?" could yield insightful responses.

While no formal ethnographies of imagination, affection, and fidelity are known to exist, other related agricultural studies show the value of this methodological approach. Kathryn Dudley's *Debt and Dispossession: Farm Loss in America's Heartland*, for example, draws on interviews and observations to describe the connections that dispossessed farmers in western Minnesota developed with their land. Her work also highlights farmers' heartbreak in leaving their land as a result of the 1980s Farm Crisis, revealing the intimate nature of farm loss. Similarly, Sonya Salamon's *Newcomers to Old Towns: Suburbanization of the Heartland* explores relationships between farmers and their rural places. She pays special attention to how these bonds are broken by sprawl and growth from nearby cities. Her case studies of "Smallville" (a persistent agrarian community threatened by urbanization) and "Prairieview" (a farming community gentrified by upscale suburbanization) are particularly relevant examples. Richard Olson and Thomas Lyson, Jillian Moroney and Rebecca Som Castellano, Michael Bell, Todd LeVasseur, and Megan Larmer, in their respective works, also show the benefits of formal ethnography by exploring farmers' human-land relationships, albeit in different ways, in different places, and for different reasons.[50]

All that to say, an ethnography that explicitly examines some farmers' persistence and stewardship motivations can help determine the presence and potential of imagination, affection, and fidelity.[51] Beyond a focus on stewardship stimuli, this approach could also reckon with the structural challenges that many rural agrarian communities in the United States face: rapidly changing social, cultural, economic, and environmental landscapes that stem from suburban sprawl, farm consolidation, and—for farmers of color—racism in agriculture. Acknowledging these ever-present systemic difficulties is critical to understanding how stewardship virtues emerge and persist.

In line with Berry's writings on people who buck the norms of modern agriculture, then, this grounded research should be anchored in communities where some farmers are simultaneously resisting temptations to expand their operations significantly or to sell their land and get out of farming completely. An analysis of stewardship motivations in this sort of setting would be a worthy and fruitful endeavor, especially if the researcher brings the passion of an amateur to the project.

While imperfect, this book tries to offer that much-needed ethnographic exploration.

Imagination, affection, and fidelity should be positioned as practical virtues. If so understood, they can be seen as essential motivators and sustainers of good stewardship. As the first of the three virtues, imagination is essential because it helps farmers understand and know their places. This grounded wisdom can then help them become more thoughtful and connected caretakers. With empathetic understanding comes affection for their places. But this affection is not temporary, nor is it sappy or sentimental. It is hard-earned and active, forceful and steadfast, akin to the love that we feel for a friend or family member that persists even through challenging times. Affection stimulates and sustains fidelity, or an enduring commitment to place—much like the commitment that characterizes marriage. In fidelity, we find genuine, long-lasting, and persevering stewardship.

In addition to positioning imagination, affection, and fidelity as virtues, it is essential to understand how these concepts operate in real life. Why engage both philosophy *and* practice? First, situating these concepts as virtues supports the notion that individuals—and not just larger entities like governments, universities, corporations, and nonprofits—can have an enduring impact in the realm of good ecological and agricultural stewardship. In other words, establishing imagination, affection, and fidelity as a collection of practical virtues shows that personal action can indeed make a difference. It shows that individuals have both power and responsibility. Virtues are serious faculties that warrant attention and development.

In the essay "Think Little," Berry writes, "A good farmer who is dealing with the problem of soil erosion on an acre of ground has a sounder grasp of that problem and *cares* more about it and is probably doing more to solve it than any bureaucrat who is talking about it in general. A [person] who is willing to undertake the discipline and the difficulty of mending [their] own ways is worth more to the conservation movement than a hundred who are insisting merely that the government and industries mend *their* ways." Here, Berry is not saying that pressing the government or industry for more responsible environmental action is futile. He has himself been arrested for protesting mountaintop removal mining in Kentucky, so he sees the value of activism and advocacy. What he is saying—and what I am trying to convey—is that without individual action and commitment, larger movements are hollow.[52] Virtues give authority and longevity to personal efforts and enlist individuals into action for more thoughtful agricultural and ecological care.

Second, establishing imagination, affection, and fidelity as virtues and testing their validity in the field could help refashion popular conceptions of some farmers as thoughtful stewards and conservationists. Of course, not all farmers fit this bill. Many, as noted earlier in this chapter, have harmed the earth through continual exploitation. But it is unfair to lump all farmers into this category, as I worry that too many people do, because many are wonderful caretakers. Some may at times hold grave views on such topics as climate change, chemical applications, and energy usage—which can certainly be seen as troubling, even dangerous, perspectives—but their commitment to place makes them important environmental allies. If they can be understood in this light, more opportunities could arise for environmentalists and agriculturalists to work together on stewardship efforts. Land conservation, for example, presents just one of the many opportunity areas for strengthening cooperation and finding literal common ground.[53]

Finally, it is worth exploring these virtues on the ground because if it's found that nurturing farmers do sustain their stewardship via imagination, affection, and fidelity, then they can be understood and uplifted as everyday exemplars that may inspire the development of these virtues in others. Their example—if supported through strategic and courageous policy—could influence other farmers to become nurturing, virtue-driven stewards who prioritize the health and well-being of communities. With exemplars to emulate, perhaps more farmers will consider good care alongside profit, not ignoring the latter but not worshiping it either. This approach to agriculture would truly serve people, places, and the planet, especially when compared with the status quo system that widely devalues or ignores stewardship virtues and, in the process, causes widespread harm. Yes, industrial agriculture is thought to be efficient, and it produces cheap food, fiber, and fuel. But at what real cost? Does production at the expense of good stewardship and husbandry equate to success? As Berry argues, without affection as a foundation in agriculture, "the nation and its economy will conquer and destroy the country."[54]

Good farmers' virtue-based stewardship could also be instructive beyond the farm, where the vast majority of Americans live. To be clear, we need more farmers in the US—millions more. New, diverse, virtue-driven agrarians could help replace the aging farmer population, as well as bring new energy and efforts to good stewardship and sustainability. With the right structures in place, an influx of farmers could also replace many of the United States' massive, industrial agricultural operations with farms of a more locally appropriate size and scale, farms that offer what Wes Jackson has called an "eyes to acres" ratio that yields the possibility of loving care. That appropriate scale, it's important to note, will differ across varying geographies and farm types. Yet not everyone should become a farmer. Encouraging a mass "back-to-the-land" movement would be unwise, irresponsible, and destructive. If a majority of urban and suburban people left their current homes and bought land in rural America, we would end up with a patchwork of tiny lots too small to support viable production and subsistence agriculture, as well as a farming populace that has no idea how to succeed in a vocation in which even experienced farmers can quickly fail. Further, a massive urban emigration to rural agricultural communities would encourage people who have not practiced or studied the art of farming to take it up, potentially leading to disastrous agricultural, ecological, and economic consequences. We need more farmers committed to nurturing stewardship virtues, and we need them to be on more small and midsized farms.[55] But we should be thoughtful about bringing that vision to life.

That clarification or caveat aside, imagination, affection, and fidelity are still relevant and essential for nonfarmers. Rather than everyone moving to agricultural settings, people could cultivate these virtues where they already are. This might mean connecting with, loving, and stewarding an urban park. It might mean paying attention to and caring for one specific tree, perhaps one planted in a nearby sidewalk or along the street. It could mean, as Berry suggests, planting vegetables in a backyard or getting more involved in a community. On top of these types of actions, perhaps it means supporting local, right-sized farmers when possible, buying food and fiber grown with care.[56]

A clear understanding of virtue-driven farmers could help establish usable exemplars for all citizens, enabling more people to localize and bolster their environmentalism through imagination, affection, and fidelity. This approach, this *work*, isn't a rosy vision. It's necessary for advancing good farming and providing better care for the earth and each other.

2

"The Hard Thing Is Keeping the Land": Reckoning with Reality

I pulled into a pasture-turned-parking-lot and stepped out of my truck. After putting on a face mask and sanitizing my hands—the protocol for in-person research during the early days of a pandemic—I started walking toward a big, white tent in the middle of the field. A few dozen people had already arrived. Some were saving seats for friends or business partners. Some were visiting the concessions trailer to buy a cold drink, already seeking relief from the hot and humid August morning. We were all there for the same reason: to watch a farm sell.

Near the tent, two men—one young, one old—leaned against the tailgate of a blue Silverado. They wore boots, jeans, and ball caps, and they talked to each other quietly as people streamed toward the tent. I nodded hello. They nodded back. "Any chance this land sells as a farm?" I asked. The younger man laughed and shook his head. Developers would buy the land, he said. They would put dozens of houses right where we were standing. "We've got a small farm about a mile away," he said, "and this shit's happening all around us." With a shrug and a sigh, he turned back toward his father, and I kept walking.

Not long after I settled into a socially distanced spot outside the tent, the auctioneer began to work. He explained the sale rules and answered questions from the crowd. With a booming drawl, he reiterated what had been advertised on signs across the county: "This land has a ton of development potential, folks. Do *not* miss this opportunity!"

The sale started, and one by one, five- and ten-acre tracts were sold. People in polos and T-shirts emblazoned with the names of construction companies threw up their hands and nodded their heads. Cries of "Yep!" exploded from the auctioneer's assistants. A few farmers tried to bid, but the price soon eclipsed what they could offer, even when they teamed up with friends and neighbors. In two hours, 188 acres were sold to the highest bidders at an average of $23,000 an acre. As predicted, developers had prevailed, their ownership soon to be marked with signatures, signs, and scars from dozers. The people selling the property—nonlocal heirs of a deceased farmer—did not attend the auction.

This scene, or variations of it, is playing out all over Robertson County, Tennessee. And it is happening more and more often. Over the past few decades, this one county has lost tens of thousands of acres of farmland. In part due to sprawl coming from the nearby cities of Nashville and Clarksville, both of which are growing at rapid rates, the future of this county's agricultural land, farmers, and rural communities is uncertain.

While the farmland loss occurring in Robertson County is acute, it's not isolated. Agricultural landscapes in the United States have been and still are being converted to residential and commercial uses at staggering rates. According to American Farmland Trust's most recent reports, roughly eleven million acres of agricultural land were converted to nonfarm uses between 2001 and 2016 alone. Due to the Great Recession, development actually slowed for several years during this period. So eleven million acres, or about two thousand acres per day, were lost despite several years of economic hardship and construction slowdowns. Forward-looking reports from American Farmland Trust suggest that unless we mend our haphazard development ways, millions more agricultural acres will be compromised by 2040. While the most extreme threats to farmland are in the South—Texas, North Carolina, Tennessee, and Georgia are ranked numbers one, two, four, and five, respectively—most every state is affected.[1]

Scene at farmland auction. (Photo by author)

In Robertson County, as well as many other communities across Tennessee, the South, and the nation, it's not just farmland loss from development that is impacting rural, agricultural communities. Once an area filled with small and midsized farms—farms that I have classified here as roughly 50–499 acres in size, which is a subjective range (and may not be appropriate for communities in other parts of the nation) but uses local context and draws on other descriptions—Robertson County's remaining farmland is steadily becoming dominated by large-scale agriculture.[2] While many of the county's biggest operations are technically "family farms" under the broad definition of the US Department of Agriculture (USDA), they encompass thousands of acres, monopolize production, and use practices and methods that are often more industrial than agricultural. These farms are extremely capital-intensive, and they often rely on complex systems of financing and credit, as well as government support, to operate. Smaller-scale farmers, many of whom also work off-the-farm jobs to make ends meet, struggle to compete. Their existence is threatened.

This combination of agricultural consolidation and rapid farmland loss leaves some locals feeling like they are getting "squeezed out" of farming altogether. For people who hope to continue caring for their farms or start new farms of their own, it tempts despair. I saw this anguish firsthand at the farm auction: in the faces of neighboring farmers who had come to watch, knowing they couldn't make competitive bids; in a young ring man who worked for the auction company, who told me he hated selling farmland but needed this job to support his own farm.

But I also saw signs of resistance from farmers and farm-service providers, land conservationists and community leaders in Robertson County. In my conversations with these folks, they discussed the challenges that small and midsized farmers face, the changes that have occurred in their communities in recent years, and the future of farming in the county and nation. While they had different perspectives on these issues, most agreed that it is hard for small timers to hang on to farmland in their community. It is not often very profitable, at least in any meaningful way. "The easy way out would have been to quit a long time ago," one woman told me. Another farmer said, "The hard thing is keeping the land. The easy thing is to sell and be done with it."

Still, many people I spoke with are making sacrifices—doing what's hard instead of what's easy—to continue caring for their farms. Using different words and phrases to describe their reasons for resilience, many unknowingly referenced Wendell Berry's virtues of imagination, affection, and fidelity as key motivators. "I just love this farm, OK?" one older farmer said. As we sat in his modest living room, just down the hall from the bedroom where he was born, he elaborated on his family's fidelity to the farm. "If somebody offered me $20,000 an acre, I'd just tell them to go to hell."

I laughed at that remark, thinking it a joke. He didn't. It wasn't.

Building on the philosophical foundations laid in chapter 1, insights, experiences, and stories from dozens of farmers and local leaders reveal vigorous, almost-visceral commitments to stewardship and their motivating factors, including those that stretch beyond Berry's virtues. In Robertson County, some determined farmers, young and old, are refusing to "get big or get out," as many government officials, extension agents, and business leaders have told them to do. These people are not expanding their farms to outcompete neighbors. They are not selling their land to become the next new subdivision or the latest addition to a ten-thousand-acre operation. Instead, they are continuing to nurture relationships with their land and place, to serve as caretakers, stewards, and "stickers." In doing so, these farmers practice imagination, affection, and fidelity on the ground.

Why Robertson County?

Before exploring small and midsized farmers' motivations to continue caring for their land, it's necessary to acknowledge the adversity they face in doing so. This recognition is especially important when understanding imagination, affection, and fidelity as virtues. Without this element of struggle and difficulty, the cultivation and honing of virtues is hollow and untested. Character is shaped during difficulty. While small and midsized farmers face various forms of adversity on a daily basis—from tough, physical, outdoor labor and tight time constraints to uncertain economic prospects and mental stress—the two major structural challenges considered in Robertson County are farmland loss from real estate development and the growing footprint of large-scale agriculture.[3]

Rapid Development and Farmland Loss

Robertson County emerged as a potential field site for this book early in the research process. I knew that even in spite of its long-established rural agrarian culture, this community was losing farmland to residential, commercial, and industrial development.

My awareness of the county's land-use change stemmed from both personal and professional experiences. I grew up on a farm roughly sixty miles south of Robertson County, and in personal travels to the county, I noticed the amount of farmland being converted to nonagricultural use. I witnessed new subdivisions popping up on former crop fields when driving to watch tractor pulls. Even on bus trips to football scrimmages and games in high school, I could see the changes in the county's literal and figurative landscape. And as noted in the preface, I worked for The Land Trust for Tennessee from 2017 to 2019. While there, I led multiple conservation projects in Robertson County. Each time I traveled to this community, it seemed that more and more farmland had been transformed.

Yet after digging into data from the United States Census of Agriculture, I realized that the rate and scale of farmland loss in Robertson County were even higher than envisioned. Among other factors, the sheer quantity of land being converted to nonagricultural uses in this community led me to focus research efforts here.

According to the 2002 and 2017 Censuses of Agriculture, Robertson County's total farmland area declined by 41,245 acres over this fifteen-year period, translating to one of the highest farmland loss rates in all of Tennessee's ninety-five counties. This time range doesn't even include more recent boom years. To provide perspective on the enormity of this figure, the statistic breaks down to a drop of roughly 2,733 acres of farmland each year in Robertson County from 2002 to 2017. That's just under 7.5 acres lost per day. And over the ten-year period between 2007 and 2017, the rate of loss was even higher.

Map of Robertson County's location in Tennessee. (Map created by Luke D. Iverson; data sources: US Census Bureau TIGER, Natural Earth, National Land Cover Database 2016, Esri Living Atlas)

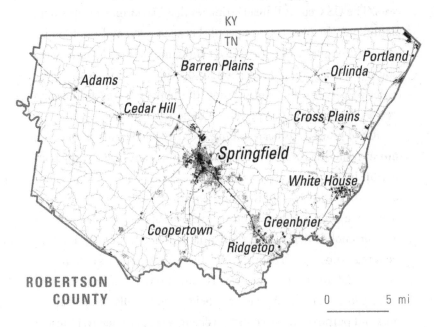

Select towns and communities within Robertson County. (Map created by Luke D. Iverson; data sources: US Census Bureau TIGER, Natural Earth, National Land Cover Database 2016, Esri Living Atlas)

Roughly 3,520 acres were lost per year, a daily decrease of about 9.6 acres. For a county with 192,000 total acres of farmland as of 2017, these figures are staggering. It should be noted that acreage statistics from the Census of Agriculture aren't perfect, partly because these figures are self-reported, lack spatial specifics, and fail to indicate what land use now occupies former farmland. Still, this data from the federal government illuminates key trends, and in Robertson County, the trend is clear. Agricultural land is vulnerable.[4]

The county's geography and proximity to major metropolitan areas explain much of its farmland loss to development. Robertson County is adjacent to both Davidson and Montgomery Counties, home to Nashville and Clarksville, respectively. Both of these counties and their major cities have experienced explosive growth in recent years. The US Census Bureau estimates that Davidson County's population grew by over 145,000 people between 2000 and 2020, an increase of roughly 25 percent. Over the same period, Montgomery County's population grew by about 85,000 people, an increase of just under 63 percent. For comparison, the United States' population grew by an estimated 17.8 percent from 2000 to 2020.

Rapid growth has been spilling over city and county boundaries into surrounding rural areas like Robertson County—where the population increased by about 33 percent over that same twenty-year range—in traditional sprawl and "leapfrog development" patterns. David Bengston and his coauthors describe sprawl as "noncontiguous, automobile dependent, residential and nonresidential development that converts and consumes relatively large amounts of farmland and natural areas." In other words, new homes, warehouses, and parking lots are replacing pastures, fields, and woodlands. For all intents and purposes, these landscape changes are permanent. Housing developments aren't torn down to sow hayfields, drill wheat, raise livestock, or plant orchards.[5]

Statistical evidence suggests that sprawl from nearby cities is more problematic in Robertson County than it is for most other rural counties in the nation. In 2014, Smart Growth America—one of the nation's leading voices for urban and regional planning—partnered with academic researchers to measure sprawl across the country. The organization released a "sprawl index" that measured development density, land-use mix, activity centering, and street accessibility in 221 metropolitan areas. Findings show that Nashville was the fifth most sprawling metropolitan area in the nation. Clarksville ranked as the country's third most sprawling metro area. Robertson County is sandwiched between these two major cities and thus feels the effects of urban and suburban sprawl from two directions. With more and more people looking to the fringes of urban areas—especially in Nashville and Davidson County, where some economic analysts believe that major property tax increases and higher costs of living will drive even more people into rural communities that surround the city—these sprawl trends are expected to continue or grow.[6]

Sprawl is also more problematic in this county than in others because of the landscape itself. Robertson County has long been a rural farming community, and its topography is generally characterized by flat and gently rolling fertile fields and scattered woodlands. For residential, commercial, and industrial developers, this landscape is a dream. When possible, developers target these types of places on the rural-urban fringe because they are relatively easy to build on. Less earthwork and site preparation need to be completed here than in areas with rough, rocky terrains, helping builders save on their expenses. Parcels of land also tend to be cheaper in rural communities than they are in urban cores or strictly suburban areas, which can lead to greater profits at resale for developers and speculators.[7] Ironically, the same physiographical factors that make Robertson County an excellent place for agriculture also amplify the loss of farmland.

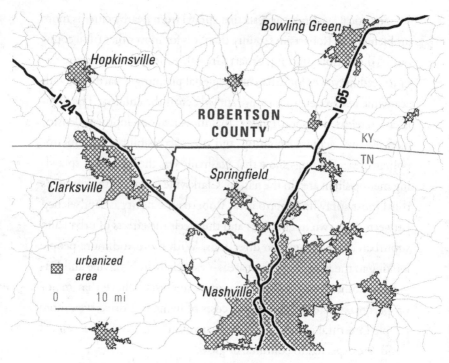

Robertson County's proximity to major urban centers. (Map created by Luke D. Iverson; data sources: US Census Bureau TIGER, Natural Earth, National Land Cover Database 2016, Esri Living Atlas)

Farmland loss from sprawl is further heightened given the ease of traveling between Robertson County and nearby urban areas. Two major transportation arteries—Interstate 65 and Interstate 24—flow through Robertson County. Interstate 65 runs along the county's eastern boundary and directly connects this part of the county with Nashville, as well as Bowling Green, Kentucky's third-largest city. Interstate 24, which connects Clarksville and Nashville, runs along Robertson County's southwestern boundary. Both I-65 and I-24 have multiple exits within or adjacent to Robertson County, enhancing highway accessibility. In half an hour, people can leave towns like

Cross Plains, White House, Orlinda, Adams, and Coopertown and be in Nashville, Bowling Green, and Clarksville.

This proximity to major cities fuels the construction of warehouses and industrial parks, several of which have been built or expanded in recent years. The easy commute also drives the development of bedroom communities, which alter the cultural, social, environmental, and economic dynamics of established communities. Whereas many smaller-scale farmers and rural people live in a local fashion, carrying on most aspects of their lives near their homes and in their towns, exurbanites who move to developments built on former farmland often only sleep there. They spend most of their lives elsewhere, usually working, eating, and shopping in the nearest major city.[8] In Robertson County and other agricultural locales, this phenomenon can dissolve tight-knit rural communities, or as Salamon writes, it can lead to the "homogenization of the vital aspects of agrarian community life [that locals] most cherish." In some ways, this development leads to rural gentrification.[9]

While sprawl is a key factor in driving farmland loss in Robertson County, other forms of development also erase agricultural acreage. In particular, scattered low-density residential development has transformed much of the rural landscape. Low-density residential development is difficult to define and varies across geographies, but it generally refers to the construction of one single-family home on a large lot. These lots are often, but not always, one to fifteen acres in size, meaning that they take up a significant amount of space for a single residence. American Farmland Trust reveals that low-density residential development has overtaken more farmland in Tennessee than has traditional sprawl. Further, American Farmland Trust explains that the simple existence of low-density development in an area heightens the likelihood that more nearby farmland will soon be converted to development. Using statistical and spatial analyses, the organization shows that Tennessee farmland near low-density residential areas is eight times more likely to be converted to nonagricultural uses than farmland that is further from these dispersed developments.[10]

In many different ways, development has consumed—and continues to consume—tremendous amounts of farmland in Robertson County. For those who are looking to sell their land, this boom is welcome. But small and midsized farmers who want to keep tending their farms—or new and beginning farmers who want to purchase land and start farming—must constantly contend with development pressures to remain or become stewards of the land.

The Growth of Big Agriculture

Like farmland loss from development, agricultural consolidation provides a challenge to small and midsized farmers. For decades, farming has been moving toward bigger and bigger scales across the country. J. D. Hanson and colleagues observe that this commercialization and industrialization is happening across farm types. Government structures and market pressures have made consolidation and concentration a major trend in dairy, livestock, and commodity crops, among other areas of agricultural production. Various news and media outlets—from local papers and radio programs to PBS documentaries, *Time* magazine, and the *Wall Street Journal*—have recently reported on the impact of these growth trends on small and midsized farmers. Due to factors largely beyond their control, people are being forced out of or struggling to enter an agricultural livelihood and lifestyle. Those who persevere face continued economic struggles. Although the data doesn't factor off-farm income into the equation, which is key for many farm families, the USDA's Economic Research Service found that, in 2019, "between 62 and 81 percent of small family farms were operating at a 'high-risk level.' "[11]

Regardless of how one chooses to measure growth, US farm sizes have risen sharply over the past several decades. On the whole, big farms keep getting bigger, and they make up a growing percentage of the nation's total agricultural production.[12] While production percentages and income data are helpful for understanding these trends, acreage statistics are particularly worthy of examination. Although imperfect, they offer an ease of understanding and on-the-ground relevance that financial and production metrics cannot provide.

According to the USDA's Economic Research Service, "cropland acreage has concentrated into fewer, but larger, farms" in recent years. "By 2012, 36 percent of all cropland [in the US] was on farms with at least 2,000 acres of cropland, up from 15 percent in 1987." A glance at the 2017 Census of Agriculture suggests that this trend has continued. Another helpful statistic for understanding movement toward larger-scale farming across the United States is midpoint acreage, which measures the point at which half of the nation's cropland is on smaller farms and half is on larger farms. According to the same Economic Research Service report—appropriately titled "Three Decades of Consolidation in U.S. Agriculture"—that number almost doubled in the twenty-five years between 1987 and 2012, rising from 650 acres to 1,201 acres. A 2014 article from the *Washington Post* puts this acreage-based discussion of farm growth in even starker terms: "As is true with much of the country's wealth, the vast majority of America's farmland is controlled by a small number of farms. The top 10 percent of farms in terms of size account for more than 70 percent of cropland in the United States; the top 2.2 percent alone takes up more than a third."[13]

Another statistic one might expect to be helpful for visualizing consolidation in agriculture is average farm size. However, the Census of Agriculture shows that the average size of US farms stayed the same—441 acres—between 2002 and 2017. Given the other data that cite a trend of increasing farm size, why does average farm size suggest that no change is occurring?

The information used to calculate average farm size leads to skewed results. To again quote analysts from the Economic Research

Service, "Means and medians do not capture the shift of acreage and production to larger farms." While large farms are indeed growing, very small farms are also becoming more numerous, which leads to little change in the average size due to offsetting extremes. In fact, with regard to the proliferation of "tiny farms," research shows that the number of farms between one and forty-nine acres in the United States increased by 28 percent between 1982 and 2012. The US Census of Agriculture shows that these trends continued in 2017.[14]

This increase in tiny farms occurs largely because the USDA's threshold for being recognized as a "farm" is low. Since 1975, the USDA has defined a farm "as any place from which $1,000 or more of agricultural products were produced and sold, or normally would have been sold, during the year."[15] This definition, which has not accounted for inflation since it was first developed, means that, for example, a person who owns a five-acre residential lot could qualify as a "farm" under current standards by producing a very small amount of agricultural products—or just showing that such production is *possible*. Of course, some people genuinely farm on very little acreage, and these operations are important to the local food movement. Specialty crop and vegetable farmers, for example, may thrive on five-, ten-, or fifteen-acre plots. It's impressive and inspiring to see what some people can produce on just a few acres of land, whether for sale or subsistence. But in reality, many tracts that qualify as "farms" are not farms at all, although they are frequently advertised as such.

In fact, one study found that in 2012, 466,000 farms in the US reported $0 in sales of agricultural products, while in 2017, nearly 604,000 farms reported sales below $1,000. Experts estimate that many of these "zero-sales" farms are owned by wealthy families, people who may be using the farm designation to qualify for a tax break. Yet some are also minority, female, and low-wealth producers who may lack the resources to actively farm their land, while some may

Sign advertising "Mini Farms" for sale near Interstate 24 in southwestern Robertson County. (Photo by author)

own open land primarily for hunting or recreation. Others could be farming for subsistence purposes only, leading to little or no sales. Still others may be farmers who didn't sell crops or livestock in a given year for whatever reason. The data contains several uncertainties, but it helps illustrate how "farms" are classified—rightly or wrongly—under the USDA's decades-old designation.[16]

The lax "farm" definition that affects average farm size across the nation certainly influences metrics in Robertson County. As noted earlier, low-density development is prominent in Tennessee and Robertson County. Large-lot residential tracts are becoming more numerous throughout much of the county, and many of these places qualify as "farms." In a nationally focused analysis, the science journalist Maggie Koerth even makes explicit mention of Tennessee as a leading state where large residential properties are frequently recognized as farms.[17]

Although the average farm size in Robertson County increased by only 16 acres between 2002 and 2017 (from 144 acres to 160 acres), large-scale agriculture is clearly becoming more dominant in this community. Some metrics calculated on a national level, such as midpoint acreage, are not publicly available for Robertson County. Other data from the Census of Agriculture, however, show the influence of big farmers in this community. For example, in 2017, the 68 farms (of 1,201 total, most of which were less than 50 acres in area) that were over 500 acres in size controlled 105,000 (55 percent) of Robertson County's 192,000 total farmland acres. Of those 105,000 acres, the 13 largest farms encompassed 52,800 acres. This means that 1 percent of farms control nearly 28 percent of the county's agricultural acreage, a stunning and alarming statistic.

The side-by-side factors of farmland loss from development and agricultural consolidation present real challenges to small and midsized farmers in Robertson County. But these dual difficulties also make this community an appropriate place to study smaller-scale farmers' perseverance. Robertson County presents a microcosm of the ways that real estate development and large-scale agriculture threaten farmland, small and midsized farmers, and rural communities across the nation.

Moving Beyond Statistics and Embracing Ethnography

The statistics and quantitative observations cited in the preceding section are important for better understanding the challenges faced by small and midsized farmers in Robertson County. But quantitative

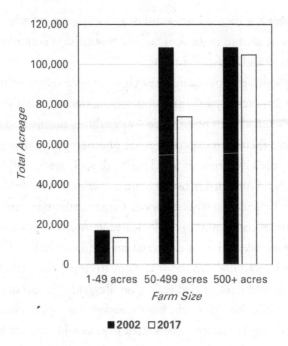

*Total acreage covered by farm size in 2002 and 2017 in
Robertson County. (Data source: US Census of Agriculture)*

data also leaves many questions unanswered. While valuable, this impersonal information cannot capture the lived experiences and perspectives of local people.

Qualitative research enables us to better grapple with the challenges faced by small and midsized farmers in Robertson County. It
can also help us explore and understand their stewardship motivations and the prevalence of imagination, affection, and fidelity. The
approaches and methods of ethnography are particularly useful. This
sociological and anthropological form of study prioritizes local perspectives and offers an intimate understanding of individuals' and
communities' thoughts, feelings, actions, and experiences. As I argue

in chapter 1, a detailed, structured, and impassioned ethnographic study is needed to make sense of the stewardship commitments of some small and midsized farmers.[18]

Typically, ethnographic research employs a variety of methods. Approaches such as participant observation, transect walks and drives, interviews, and even photovoice—or enabling people to document and share their own experiences via photography—offer glimpses into people's experiences and lead to deeper, more well-rounded knowledge. I intended to use all of these methods in my field research in 2020. I planned to work alongside farmers, offering them my labor as a form of compensation while learning more about their actions and experiences through conversation and observation. Given my farming background, free labor was not an empty compensatory gesture. It would have carried real value. Beyond this approach, I was going to walk through fields, woods, and pastures with willing farmers, seeing the landscapes where their lives unfold, and drive around the county with locals to see and hear about their community and culture. I even intended to give interested farmers disposable cameras so they could document and share their perspectives via imagery.

But because of the COVID-19 pandemic and the public-health measures that were needed to protect myself and others, these plans for ethnographic field work changed. Except for taking many drives throughout Robertson County alone, attending one farmland auction, and conducting in-depth interviews—most of which were done over the phone—planned research methods had to be shelved. While participant observation, guided transect walks and drives, and photovoice would have provided in-depth insights into the experiences of people in Robertson County, these methods were unsafe because of COVID-19.

Despite the methodological challenges presented by a pandemic, I still conducted ethnographic research. And while the research was constrained, it proved fruitful. I interviewed thirty-one people who lived or worked in Robertson County. These research participants fit roughly into four groups. Most (nineteen) are classified as small and midsized farmers. With the exception of three farmers located on or less than one mile from the county line, all lived and farmed in Robertson County. A smaller number of people (five) are labeled as community leaders, who range from elected officials and a town manager to a county planner and an industrial advocate. In addition to their roles as community leaders, a few of these folks also live on and care for small farms, so they provide multiple perspectives. Beyond farmers and community leaders, I spoke with individuals classified as farm-service providers (four) and professional land conservationists (three). The service providers interviewed assist Robertson County farmers with soil conservation, agricultural crime prevention, and farm planning. The professional land conservationists work for two different nonprofit organizations headquartered in Nashville that are actively working to conserve farmland, advocate for smart planning, and serve communities in Robertson County and beyond.

Connecting with people for interviews was a challenge because of COVID-19. Under normal circumstances, I would have met new people in person in friendly, local settings like farm stores and diners, but given the pandemic, I relied on four primary approaches for connecting with participants. First, I identified people, such as farm-service providers and community leaders, via online research. I located email addresses and phone numbers on government and university websites and then "cold called" folks. Roughly half of the people I contacted via this method agreed to speak with me after learning about my research, a somewhat surprisingly high figure.

Second, I worked through a list of farmers provided to me by the Tennessee Century Farms Program, which is anchored at the Middle Tennessee State University Center for Historic Preservation. Century

Farms are recognized as places that have been actively farmed and owned by members of the same family for over one hundred years, which means that every person on the list has a long connection with their land. Many of the phone numbers in the provided form were disconnected, and few entries had associated email addresses, but the contact list provided by the helpful staff at the Center for Historic Preservation led to multiple interviews.

Third, I contacted farmers in or directly adjacent to Robertson County who had worked with a land trust to protect their land from development. Excited about my research and hopeful that it would serve rural communities, leaders at this organization kindly agreed to provide contact information for these farmers. It is important to note that I had met some of these farmers in my two-year stint working with this conservation organization, so they did not perceive me as a stranger.

Finally, I used "snowball sampling" to connect with other people. Some farmers and community leaders connected me with other interested and willing interviewees who in turn referred me to others. Their help was essential. While there may be elements of bias inherent in these sampling methods, the pandemic left few other options for connecting with people.

A major shortcoming of the research in Robertson County lies within the demographics of participants. While there is gender diversity among the people interviewed for this case study, there is almost no racial diversity. Only one person is nonwhite: a Black farm-service provider. Despite ardent efforts, connecting with people of color in Robertson County's farming community was an immense challenge, largely due to the lack of diversity within the county's agricultural sphere. According to the 2017 US Census of Agriculture, there were 1,991 total "producers," or farm owners, operators, and tenants, in Robertson County, but only 30 of these producers (1.5 percent) identified as nonwhite or as "more than one race." Of these nonwhite producers, only 11 were Black farmers. Through conversations with farm-service providers, as well as a Black historian in the county, I learned that a few of these 11 Black producers had died of old age since the Census of Agriculture was taken. Another had grown unable to continue caring for the farm and sold her land. Even when it seemed that I had made a breakthrough thanks to a local leader who provided contact information for two Black farming families, I was ultimately unsuccessful. The phone numbers shared were landlines that had been disconnected.

Listening to and learning from the perspectives and experiences of Black farmers was a goal of my research from its earliest stages. Given the perseverance and commitment required of Black farmers to endure individual and systemic race-based challenges, as well as farmland loss, consolidation, and the everyday difficulties of farming, their thoughts on stewardship virtues are especially meaningful. Thus, because of the inherent challenges of connecting with Black farmers in Robertson County, I branched out into a new community with more agricultural racial diversity toward the end of my field work. Chapters 5 and 6 highlight the experiences of Black farmers in Maury County, Tennessee, a community south of Nashville that has been affected by farmland loss and agricultural consolidation in ways comparable to Robertson County.

With regard to the interviews with Robertson County participants, I conducted five in person, adhering strictly to public-health measures such as wearing a two-layered face mask, avoiding close contact, and sanitizing hands and surfaces frequently. The remainder of the interviews were conducted via video or phone calls. Because reliable internet access is sparse in rural Robertson County—and because all farmers expressed a preference for this option if meeting in person wasn't possible—most remote interviews were conducted by phone.[19]

Before every conversation, usually days in advance, interviewees were provided a consent form to read that explained the project and their rights as participants. Each person verbally agreed to participate before their interview began.

Whether in a masked face-to-face setting or remote, all interviews were semistructured, which allowed conversations to remain focused while still being informal and relaxed. Unlike rigid, structured interviews—which too often take the form of verbal questionnaires—this approach left room for elaboration and follow-up questions while allowing individual personalities to shine. For these interviews, I followed one of four interview protocols—one for each category of participant—to limit bias and ensure consistency. When circumstances demanded, I refined or made slight alterations to these interview protocols. For example, after conducting the first few interviews, I realized that one question was confusing, so I reworked and reworded it. Further, when a person fit into two different categories—such as a community leader who also cares for a small farm—I merged questions from two protocols to better address dual perspectives.

In all these conversations, I did my best to make people feel comfortable, especially given pandemic circumstances. This commitment led me to incorporate myself into interviews, perhaps more than I would have in a normal situation. At times, I referenced my own farm upbringing and in-depth understanding of agriculture. I used local words and terms rather than academic jargon. I talked about my favorite cattle breeds and joked about the many faults of John Deere tractors compared to Internationals. I thanked people when they were vulnerable and comforted them when they cried. I did not hide my sometimes-thick accent, which mirrored theirs.

Because I acted as an empathetic person instead of an indifferent investigator, farmers and others opened up and shared raw, personal, and moving thoughts and stories. As far as I can tell, people were honest and forthcoming. While I do not think that inserting myself into the interview process caused anyone to answer questions in inauthentic ways, it's important to recognize my role in these conversations.

Once I had finished transcribing all the interviews—to again reference my southern accent, automatic transcription software struggles to accurately capture my and others' words, so this work was done manually—I began to analyze content. Initially, I planned to print each transcript and mark up the documents by hand, but with over four hundred single-spaced pages of transcriptions, I shifted strategies. Ultimately, I used a computer program to analyze interviews. I created distinct codes for topics or themes that were regularly mentioned or were a focus of my research. Then, I carefully sorted quotes and context from each interview transcript into these codes. This system made the interpretation of large amounts of qualitative data orderly and efficient, or as efficient as reasonably possible. This process still took an awfully long time.

As mentioned elsewhere, the interviews just described form the heart of my research. In chapters 3 and 4, I rely on locals' lived experiences to better understand the Robertson County farming community. These rich personal contributions address the book's central goals in compelling ways.

3

Farmland Loss and Big Agriculture:
Firsthand Accounts of Adversity

Small and midsized farmers know firsthand the discomforts of agricultural life, including at times the very work of farming itself. Almost all the farmers I interviewed spoke about the demanding physical labor required of this vocation. Feeding cattle in the snow, fixing fences in the heat, running combines until midnight, checking on pigs in freezing temperatures, cutting tobacco in the blistering sun, hauling hay alone, milking cattle every day of the year, scrambling to fix broken machinery: these were all jobs and tasks that farmers described as taxing. Some explained that after decades of farming, this work has taken a toll on their bodies. Carl, an older farmer near White House, noted the multiple medical procedures he has needed to put his body back together from farm injuries and daily wear and tear.[1] "Pain is just something I deal with every single day," he said. "Every single day."

Financial challenges for smaller-scale farmers are real too—especially for those who don't yet own their land outright and are trying to make annual payments. Farmers noted the rising costs of feed and fertilizer. They discussed the razor-thin margins that are now commonplace, whether in raising livestock, milking cows, or growing corn, beans, produce, tree crops, or other products. They also spoke about the expense of purchasing and maintaining equipment. Even nonfarmers made mention of these exorbitant infrastructure costs, where implements, tractors, and combines can cost tens—or even hundreds—of thousands of dollars. These economic factors are part of the reason why many small and midsized farmers struggle to break even some years, and they shed light on the financial barriers to entry for new and beginning farmers. These factors also explain why so many farmers must work off-farm jobs to generate income.

While these difficulties are worthy of discussion—and certainly worthy of further research—they take a backseat in this book to the issues of farmland loss from real estate development and the expansion of large-scale agricultural operations. Farmers, community leaders, farm-service providers, and land conservationists all spoke to the prevalence of these challenges.

"You Can't Throw a Rock from This Farm and Not Hit a House": Bearing Witness to the Transformation of Farmland and Rural Communities

When people were asked if they had noticed changes in the Robertson County landscape in the past fifteen to twenty years, they responded almost unanimously. Tremendous changes have occurred, they said, and former farms have been replaced by residential and commercial constructions. Brian, who owns and cares for a small farm near Adams, gave a representative answer when asked about noticing farmland loss in his community: "Good Lord, yes! [The county] has changed drastically." One farmer—Denise, who works a full-time job in a nearby city and helps tend her family's farm—went so far as to say that, compared to the rural areas they once were, parts of the county are now unrecognizable.

While many people noted the increasing reach of development over longer time frames—for example, describing the changes that have happened in their lifetimes or in the past few decades—several also mentioned the immediate, present-day erasure of farmland for development purposes. Frank, who raises beef cattle and hay just east of the Robertson County line in Sumner County, mentioned that he constantly sees farmland being bought up by developers. "More and more every day. Every day," he said. "I mean, it's just unbelievable." Frank explained that he and his wife get frequent phone calls asking if they would be willing to subdivide and sell their land. "You'd be surprised how many people will call here in a month wanting to buy our land. We've got 'em calling out of California, Arkansas, and everywhere. I don't know how they find out about it, that we even got a place. But they sure do."

Frank is not the only one to receive unsolicited offers from interested buyers, nor is he unique in recognizing the landscape changes in his community. While some people initially expressed surprise at the county's decline of over forty-one thousand acres of agricultural land between 2002 and 2017, nearly everyone agreed that the figure seemed plausible and accurate. "I wouldn't be surprised if it was actually even higher," said Nathan, a land-conservation professional.

Officials within the Robertson County Planning and Zoning Department are also attuned to the increasing growth and development consuming open land across the county. Fielding frequent rezoning requests and planning questions, they are intimately aware of the farmland conversion happening around them. A senior official and part-time farmer named Logan observed that farmland loss happens "on a weekly basis, perhaps. It appears that the family farm is going away." Elected leaders and farm-service providers have observed these changes, too.

Affirming the quantitative data cited in chapter 2, people shared that development occurs in two primary forms: industrial/commercial and residential. Industrial and commercial development occurs less frequently in Robertson County, they explained, but when it happens, it takes up large amounts of space. Frances, who helps recruit businesses to Robertson County in hopes of creating jobs—some of which go to part-time farmers to help support their farms and families—mentioned a major industrial park built on roughly one thousand acres of farmland in the early 2000s. Local, national, and international companies filled this space, aptly named the Tennessee-Kentucky Industrial Park because of its location near the state line. Other industrial parks, warehouses, and distribution centers have also been built or expanded throughout the county in recent years. Often, these commercial and industrial areas are accounted for in Robertson County's 2040 Comprehensive Growth Plan.

This forward-thinking planning approach is not always used, however. A farm-service provider and part-time farmer named Henry told an illustrative story. His neighbor's sixty-five-acre farm was passed on to heirs and then recently sold. The heirs were not interested in farming. But before officially buying the land, the new owner wanted the property rezoned for commercial and industrial use. He promised new industry and an "upgraded water line" for rural citizens, something Henry said was unnecessary and unwanted. Although the rezoning motion originally failed, industrial leaders in Robertson County and nearby Nashville guided the measure through on a second attempt in the name of creating new jobs for the community. The land then sold. According to Henry, the new owner immediately put the land back on the market at a price almost twice as high as what he had paid, hoping to cash in on the zoning change. Henry was angry that taxpayer-supported industrial leaders helped usher in a zoning change that will enhance one person's wealth while changing the community in a way that he—and, according to him, other community members—sees as invasive.

Evidence shows that other recent rezoning measures, especially for residential purposes, have also left longtime rural residents feeling cheated or ignored. In a 2017 newspaper article, for example, one citizen commented on yet another new residential subdivision proposal in the community: "This is what urban sprawl looks like. It sucks the life out of your towns. Where does it stop? We'll be right back here in a few months with a request to rezone the land touching this and across the road from this. You can only lose your integrity and character once. You don't get it back." Her comments point toward data from American Farmland Trust that suggest that real estate development in a rural area makes additional development nearby much more likely.[2]

As seen earlier through Frances's and Henry's comments, industrial and commercial development drives a portion of farmland loss in Robertson County. But according to locals, it is residential development that takes the biggest toll. "Oh, it's houses," said Denise when asked what is replacing farmland in her community. "Traditionally, Robertson County was a farming community. We're losing those values. We're just becoming a suburb of Nashville. We're losing that small-town way of life. I grew up in a four-way-stop town. And now, the traffic going through here is hard to fathom, compared to what it was years ago."

In line with the data cited in chapter 2, the housing that Denise mentioned comes in the form of both traditional sprawl and large-lot residential development. Interviewees shared that an abundance of subdivisions—some large, some small—have been built in Robertson County over the past two decades. "This is just subdivision city," sighed Phillip. He was speaking of Cross Plains, the small town near his family's Century Farm. "It used to be a rural community, agricultural, and now it's just, you know . . . Houses have growed up." Brian, a farmer near Adams, shared this perspective. "Today, as we speak, I can carry you and show you half a dozen new developments going on, from thirty to forty acres to seven hundred acres. It's been a drastic change," he said.

Drives through the county confirm these comments. While farmland conversion statistics and audible descriptions are shocking, it's even more eye-opening to physically see entire farms carved up with bulldozers, being prepared—as Phillip said—to grow houses. From highways, byways, and one-lane roads, I saw this rural community changing.

A high-level county elected leader and small farmer named Kyle tried to explain why these subdivision developments happen: "These developers learn they can buy these farms . . . and put sewer on them, tract [i.e., divide] them up into acre lots or half-acre lots or whatever they can get by with." Some, he said, quadruple their money once they resell, giving more power and wealth to those who already have it. William echoed Phillip's comments. "Sometimes a developer will come in and buy the whole farm, and he'll develop the whole farm, just build a subdivision in there. Maybe put forty to fifty houses on a spot," said the longtime farmer. Citing information they are privy to given their occupations, Kyle, Logan, and John—a rural town manager—mentioned a handful of subdivisions currently being built or planned that will add nearly one thousand homes in the eastern part of Robertson County once complete.

Likewise, low-density residential development is also transforming farmland in Robertson County communities. Some folks argued that this type of growth—and the farmland loss that accompanies it when not planned accordingly—is even more pervasive than sprawling subdivisions, an observation that mirrors statistics from American Farmland Trust.[3] "There's a lot of five-acre lots that I've seen," said Sarah, a cattle farmer who lives southwest of the county seat of Springfield, near Coopertown. "Around me immediately, it's five-acre tracts. Down the road some, it's way more concentrated." A farm-service provider named Benjamin noted the same trends. Speaking of Robertson County as a whole, he said, "Most of our land that's taken out here is primarily for personal use: building a home, whatever. [It's] five-, ten-, fifteen-acre plots that our land is being sold in. . . . We have land that's coming out of production that's being developed into smaller tracts for personal use, nonagricultural use."

Others also commented on the increase of large residential tracts on former farmland. "It's primarily single-family houses," noted Frances. "We've seen a lot of big tracts that get divided up into five acres, so people are building houses with a pretty big piece of property around it, which is—I mean, that's nice, but it's not the most efficient use of land, for sure." Melissa, a longtime land conservationist who has

New subdivision under construction in eastern Robertson County.
A neighboring farmer's barn can be seen in the background in the bottom photo.
Quantitative and qualitative data show that this farmer will be impacted by
this new subdivision and that their land is now more likely to be developed.
(Photos by author)

*Sign advertising five-acre tracts for sale on a
current farm in central Robertson County. Similar advertisements
can be found throughout the county. It's almost certain that crops
won't grow here much longer. (Photo by author)*

worked on more than a dozen projects in Robertson County, made
the same observation. From her perspective, "It's not so much traditional residential subdivisions or anything but just those home lots."

Melissa also made specific mention of "little five-acre farmettes."
Others commented on this tiny-farm phenomenon too. Sitting with

Relatively new low-density development on former farmland in northeastern Robertson County. (Photo by author)

her husband, Randy, as we spoke in the living room of their farm-house, Janice observed, "So many of these farms have gone into these mini-farms, these five-acre lots. You see that all over the place." Developers specifically advertise how these lots can be carved out of larger, once-agricultural land areas. They encourage people to build "Barn-dominiums," a play on "condominiums," and bring suburbia to rural landscapes. Advertisements and signs across the county—as well as on real estate websites publicized in nearby cities—reveal that the proliferation of miniature "farms" and low-density housing development affects numerous Robertson County communities.

As nearly every person I interviewed shared, the explosion in large-lot residential development and housing subdivisions is reshaping this once-rural community, transforming its locales into what a few people called "bedroom towns." They mentioned some of the same issues with bedroom towns that are noted in earlier-cited studies: lack of time, energy, and financial investments in the local community, as well as environmental and agricultural damage.[4] But rather than just acknowledging the occurrence of farmland loss in various forms, it is important to note where within the county most loss is occurring. Raw farmland-loss statistics do not contain this information, so qualitative data from locals takes on even more significance.

Participants shared that the most extreme losses of agricultural acreage to development are in the southern, southwestern, and eastern parts of Robertson County. They specifically noted that areas around White House, Greenbrier, Ridgetop, Coopertown, Cross Plains, and Orlinda are experiencing farmland conversion. In some ways, this geographical distribution isn't surprising. These towns and their outlying areas are all near an interstate highway, making it easy to commute to Nashville or Bowling Green or Clarksville and thus drawing people who work in these cities to the county. According to Kyle, Nashville and Clarksville especially are both branching out into rural areas. Joyce, who helps farmers find resources to hang on to their land, shared the same sentiment.

Others living in these areas also believe that these parts of the county are seeing the most growth. A comment from William, whose farm is near Cross Plains, is worth quoting at length. Rationalizing the increased development near his farm, where new homes are "as close to [him] right now as they can get, . . . right up on [his] border," he said,

> We're close to Interstate 65. A lot of people commute to Nashville. You get on the interstate, and it's a hop, skip, and a jump up here from Nashville. They don't want to live in Nashville. They want to

get out here and live in the country. That's one of the things, I think, [that] is [driving] demand for the property because people are trying to get out of Nashville or out of the big cities. They want to live out in a rural area, but they want to work in the big city. And I think the commute—we're probably a twenty-five-minute commute from here to Nashville, on a busy day probably thirty-five minutes to downtown Nashville.

His nearby peers, as well as those in other parts of the county, largely agreed. Some went as far as to say that their rural communities are quickly becoming "North Nashville" and "East Clarksville." When so many people move to the country and sacrifice farmland in the process, it isn't really the "country" anymore.

Development is also occurring in the center of the county, mostly around its biggest city: Springfield. From the core of this city, real estate development has fanned out and, in some cases, even connected with the aforementioned towns near Interstates 24 and 65. Bradley, whose farm is in Kentucky but borders Robertson County, said that the area between Springfield and Greenbrier, for example, is now almost "solid businesses and houses." It was once filled with small farms, people raising tobacco, livestock, hay, and more. Even without an interstate running through the city, Springfield and its surrounding areas have seen their share of farmland conversion.

The northwestern and north-central parts of the county have, according to several individuals, experienced far less farmland loss than other areas have. Some low-density residential development has occurred—a chunk of which is the children of farmers building new homes on family land, folks shared—but because these areas are less accessible to major cities, their farmland has been somewhat spared. Limited infrastructure also slows growth, shared Kyle, who lives near Adams. Small water lines and the lack of sewers impede developers. "I thank God every day that it's a small line," Kyle shared, speaking of the water line in his community, "because it will limit growth in that end of the county." For these infra-structure-based and geographical reasons, three interviewees who live and spend most of their time in these parts of the county explained that farmland loss to development is not a huge concern for them personally, and it is not something they have witnessed on a large scale. To them, what is more concerning in these areas is the expansion of large-scale agriculture, which will be discussed in the next section.

In the same way that ethnographic data clarifies the distribution of farmland loss in Robertson County, it also helps illuminate why this loss is occurring. Clearly, quantitative and qualitative data show that residential and commercial land is in high demand in Robertson County. But the demand element of farmland loss is only part of the puzzle, albeit a critical one. It is also important to consider the supply side of farmland loss and understand why so many people are selling their land to developers.

Locals shared a number of reasons and theories for why farmland owners in Robertson County are letting go of their land. Some, they said, are tempted by the high prices developers and urbanites are willing to pay, especially in the fastest-developing parts of the county. Multiple farmers described being offered over $20,000 an acre for all or part of their farms in recent years. I also observed these prices myself at the auction described in chapter 2, and they're reflected in online listings across various real estate websites. For someone who owns 100 acres, that price per acre leads to a $2 million payday. Charlotte, who is a town mayor and owner of a 120-acre farm, noted the allure of these prices. "I have no intention of selling my farm," she said, "but if somebody came by and made me an offer like that . . . I might have to back up and think about it." In the past, her children put this situation in blunter terms. "My kids said, 'Mom, if you get offered that and you don't sell, we're having you committed [to a mental health facility].' "

Understandably, the pull factor of high land prices is at play in Robertson County, and as seen in Charlotte's comments, it is especially prevalent when land passes from one generation to the next. In many cases, "the kids don't care nothing about farming," mentioned Kyle. Many others agreed. Logan carried this observation further, saying, "The younger generation—the millennials, if you will—is not interested in farming. . . . So a lot of the land that has been in their families for years . . . Granddaddy's passed on, Dad's not in good health, so they'd rather sell the land or a bulk of the land to someone else." Citing the formal concept of "generational turnover," the land conservationists Melissa and Nathan also spoke to this catalyst in farmland loss.[5] Generational turnover is a constant concern—and sometimes a motivating factor—for farmers who are considering permanently protecting their land from development with a conservation easement.

While Kyle's and Logan's experiences are valid and grounded in their own empirical observations, these comments perhaps paint the issue too simply. After all, passing a farm from one generation to the next presents complex estate-planning and succession issues, and for farmers with more than one child, issues of treating heirs equally can complicate the desire to conserve family land. If a farmer has four children and eighty acres, for example, should she leave twenty acres to each child, splitting up the land evenly but effectively ending its capacity for continued agricultural production? Or should she leave the land to the child who has expressed the greatest interest in and commitment to the farm and try to find different ways to support the other heirs? What if she doesn't have any real assets to leave the others? Other questions abound, too. Should this farmer continue farming herself for as long as she can—perhaps until she is in her sixties, seventies, or eighties—or should she transfer the land decades earlier so that one or more of her children can have a go at running the farm while still young and before settling elsewhere? If she chooses the latter option and has little money saved for retirement—a common situation for many farmers—what should she do in her final decades? Where should she live? If she has no children who are interested in farming or no heirs in general, should she try to find someone who is not a family member to serve as an apprentice of sorts? If so, how?

For farmers with little wealth other than the land itself—for people who are sometimes referred to as "land rich and cash poor"—these sorts of questions, which are being asked daily with real consequences in rural communities across America, are even more difficult to answer. These not-so-hypothetical scenarios make it easy to understand why some heirs feel that they have little choice but to sell family land. There are strategies that farmers, heirs, and supporters can use to conserve land in these scenarios, but even when planned with care, generational transfer and farm succession are difficult, emotional, and complicated.

Transition issues aside, other challenges affect the continuation of family farmland. Some heirs deeply want to hang on to their family's farm, but hanging on can be hard. As has already been noted, farm work is difficult, and on small and midsized scales, it is often not lucrative. When coupled with the expenses of maintaining the farm, even paying the annual property taxes can present a significant challenge for many farm owners. This difficulty leads people into financial struggle or, as so many participants described, working multiple jobs just to keep their farm, which can exacerbate physical and mental exhaustion. The struggle of hanging on is even more difficult for people who don't inherit land or who inherit land with debt still owed and must make annual farm payments on top of everything else. "If you're in debt up to your eyeballs, that puts a lot of pressure on," said William. That pressure can lead farmers to sell their land. It can also translate into more devastating results, said William. In a somber tone, he noted the recent rise in farmer suicides, a heartbreaking phenomenon felt in his own community. Some figures show that farmer suicide happens at higher rates now than during the peak of the 1980s Farm Crisis.[6]

The narrative of "the heirs don't care about farming" can be accurate. It's true that some, perhaps even many, children of farmers have every intention to sell land once they have the opportunity. Several farmers, farm-service providers, community leaders, and land conservationists noted this dynamic. But they also highlighted the difficulty of landownership for new and established farmers alike, which has certainly led to farmland conversion to development in Robertson County. They mentioned the near impossibility of new and beginning farmers competing with developers or investors to buy a plot of land as well.

Other factors—such as the logistical and cultural difficulties of farming in suburbia, which were noted by multiple individuals—spur farmland loss too. Some farmers, for example, mentioned the anger their new neighbors showed when having to drive behind slow tractors, wait on cattle to cross the road, or smell manure from livestock. This anger sparks frustration. At times, farmers even feel unwelcome in their own communities, places they've lived for generations. Greg mentioned that when a piece of his hay equipment broke down on a weekend, he had nowhere to turn for help. Ten years ago, he would have called one of his neighbors and asked to borrow a part or even an entire implement. With no more farming neighbors, he said, that option is gone. And with fewer farmers around, the need for local farm stores and repair businesses is diminished, illustrating the domino effect of farm loss in rural communities.

So far, comments and stories from local people have helped deepen understandings of what kinds of development are consuming farmland, where this development is occurring, and why it is happening. A final ethnographic insight into farmland loss concerns the phenomenon's emotional impact. Scholarly studies often neglect to include this analysis, perhaps thinking it irrelevant or insignificant. But for the purposes of this project—specifically, understanding the multilayered difficulties farmers must overcome if they hope to continue practicing stewardship virtues and tending their land—grappling with the emotional toll of disappearing farms is essential.[7]

Some participants, particularly those who do not farm themselves, expressed their emotions in general terms. Seeing a farm carved up for a subdivision made them feel sad. Others were disappointed. They didn't like that Robertson County is losing its rural agrarian culture, but their emotions were subdued. "Having seen it happen so much, I know that you win some, you lose some," explained Joyce. William's emotional response was perhaps the most nonchalant of all: "Well, if I was young, it would probably really upset me because I would know what the future would be. But I'm old now, and it's probably not gonna affect me too much one way or the other. I mean, at my age [seventy-seven]—another forty to fifty years is all I'm going to live!" he joked. What happens on the ground won't matter much to William when he's buried under it.

But the majority of interviewees—and especially farmers—experienced the emotions of farmland loss in their community on an acute level.[8] A common feeling described was "heartbreak," an emotion typically reserved for expressing deep and profound pain. "For me personally, it's heartbreaking," John said. "I think that's true for a lot of [my neighbors]. It really is. It's just depressing." Sarah echoed these sentiments. "A farm down the road from me, when the father died, the kids broke it up and sold it for housing developments. It kills me. I go through these fields and these beautiful places that used to be fields with crops and cattle, and now they're housing developments. . . . It's so sad to me. So much of it in our area—the area I live in is the Flewellyn and Coopertown area, and it has always been very, very rural. It's heartbreaking to see it happening."

Beyond heartbreak, many others used the terms "hurt" and "hurtful" to explain their emotions. Some described "hating" to see farmland developed and "grieving" when it happens. Denise felt "devastated" by the changes around her. Robert, a longtime Cross Plains farmer, said it made him "sick" and that he wished he was a billionaire so he could buy up land and slow development. "You don't really know whether to cuss or cry, one or the other," said Pete, who grew up on a farm in Robertson County and now farms just beyond the border in Kentucky. "It's sad. Because I can look at fields that have houses on them now that, when I was younger, were being farmed. But what do you do?" His rhetorical question at the end was echoed by others, revealing a shared sense of exasperation.

Justin—whose family has been farming the same land near Adams since the late nineteenth century—explained that his emotional reaction to farmland loss even manifests itself in his subconscious:

> I'll tell you this. I used to have a recurring nightmare about it. I'm not kidding. I used to have a nightmare where I would wake up and walk out of the house, and it's completely surrounded [by houses], or there's a massive mall, or there's a Burger King and a Krystal and a Walgreens all sitting at the end of my driveway. I still have that dream occasionally. Then I'll wake up. I'm not exactly sure what that's about. It's definitely doing a number on my subconscious. . . . I think the people that buy these places and live in these places, I don't think that they think about the impact. I really don't.

The permanence of the changes occurring around the county is one of the factors that bothers Justin most. Others felt the same way. Nathan observed, "Once farmland is lost, it's lost forever. Just seeing a property change hands and become developed for any sort of future nonfarm use means that it will likely not go back to agriculture."

While these emotional reactions illuminate the psychological toll that farmland loss and rapid community transformation can take on rural people, the strongest feelings were expressed by farmers who have, for whatever reason, lost some or part of their own land. Brian shared a unique perspective. His family—grandparents, aunts and uncles, and parents—lost a large hunk of their farms in the Farm Crisis of the late 1970s and early 1980s.[9] About seeing this unfold before his eyes, Brian said, "From a farmer's perspective, it's sad. As you know, farming is a way of life, and to see farm family units that worked for generations to have that kind of lifestyle, to see that go away . . . I see a lot of things that have been lost because of that. Namely, family togetherness and community togetherness—those kinds of things. Family and community unity." Having become hardened—or "callused," he said—by the experience, Brian has actually turned to working as a part-time real estate agent himself. While he doesn't like to cut up farmland, he "kind of took the attitude that if somebody is going to do it, [he] might as well be the one doing it. And that money helps sustain [his] farming operation."

Others expressed rawer emotions. Carl and his family had to sell some of their farm's road frontage a few years ago. Carl's parents both ended up with Alzheimer's at the same time, and they needed specialized care for about a decade. After a few years, Carl was hurting financially and had to make the painful decision to sell some of the family land for residential development so that he could care for his parents. With a few nearby neighbors selling land too, Carl said, "You can't throw a rock from this farm without hitting a house." When asked about the experience, he said it was "traumatic" for him. "You have to . . .," he trailed off, voice cracking. "It's . . . I knew this was a question you were probably going to ask me, and I thought I was prepared to answer it until right this very moment." It was particularly hard to discuss, Carl said, because he knew that every generation before him could have made the decision to sell this land, but they didn't. Carl "did the best he could do with the facts he had" and believes he made the right choice. His parents needed care, and he was able to provide that. Still, the experience haunts him.

Greg, who has a ninety-five-acre farm outside White House, told a similar story. All the farms around him have been converted to

New homes being built near White House on former farmland, a site that stirs painful emotions for some farmers. (Photo by author)

nonagricultural uses, typically after his neighbors—some of whom were close friends—passed away and left their land to heirs. "I am surrounded by land that has been sold here in the last three to four years. And it's a pretty substantial amount of property that has been sold," he said. "How does that make you feel?" I asked in our interview. There was a long pause, and I was afraid that our call had been disconnected because of poor cell service. But after twenty seconds or so, Greg stammered, "Just a minute . . ." He was crying. Trying to collect himself, he apologized. "I'm sorry about that. . . . Obviously, I get pretty emotional about it." His voice cracked again. "It's an eye opener how much land is going away from farming. Seeing that much around me, it's just . . . It's something I didn't think I'd see in my lifetime." Some of that land going away from farming was his own family's. At the beginning of the 2000s, Greg and his wife sold twelve acres of road frontage. Reflecting on the experience, he said, "It was hard to do. But we needed money. It was tough."

These diverse emotional reactions reveal the intimate nature of farmland loss, which cannot be captured by numbers alone. Conversations with local community members show the pain that can accompany rapidly changing landscapes and highlight the adversity farmers face when trying to persist in place. These personal reflections also emphasize the role that stewardship virtues play in powering perseverance.

"The Little Farmers Can't Keep Up with the Big Ones": The Growing Footprint of Large-Scale Agriculture

Quantitative data and statistics show that farming is trending toward bigger and bigger scales across the United States. The same is happening in Robertson County, putting small and midsized farmers on the margins. Participants acknowledged these changes and trends on the ground by offering personal stories and observations of agricultural consolidation. While only a few are quoted here, roughly 75 percent of people I interviewed spoke at length about the proliferation of big farms in the past few decades and especially in recent years.

"I hate it," said Kyle, "but the farm program we've got today put the little guy out of business. It did. The little farmers can't keep up with the big ones. You either had to get bigger or get out." Throughout the county—and especially in the northern and western areas—this "get bigger" approach, which has been fueled by government and corporate forces, has encouraged a size explosion among a few farms and an erosion of many small and midsized farms. While a fifty-acre farm was a "good-sized farm" a few decades ago near Adams, Janice and her husband, Randy, have seen those "smaller farms become parts of larger farms." In interviews, some people named farmers who are managing operations that are now thousands of acres in size, saying it is standard for one person to control five thousand acres—ten times larger than the biggest midsized farmer who was interviewed for this research—or more. A few farmers in the area manage over ten thousand acres. For Tennessee, that sort of acreage is massive.

Speaking from a small farmer's and community leader's perspective, Kyle observed that agricultural consolidation "is probably the biggest problem" the community is facing. "Small farms are disappearing. It's because they can't afford to make a living off a small farm. . . . So land is sold to whoever can pull in there with five combines. That's where I see the biggest change. The family farms are leaving because they can't afford to work it." It is worth noting that five combines would easily cost millions of dollars, which highlights the capital-intensive nature of large-scale agriculture.

According to some participants, these massive farmers are not really farmers at all. Instead, they are "managers." While Bradley was describing recent changes in agriculture and speaking about how much he enjoys the actual physical work of farming, he said, "The biggest difference I see is that your farmers aren't farming. They're managing. And that's one reason I've never become a big farmer. I like doing the work myself." Similarly, Frank noted that when farms get to be thousands of acres in size, the work becomes more like business management than farming. And to again quote Kyle, who expressed some of the strongest feelings about large-scale farming of any interviewee, "They're managers. Everything is on a computer. Put the tractor out in the field and pay somebody to get on it. At the end of the day, [the manager] figures up what his labor [cost] was." Managers need more business skills than agricultural acumen.

No matter how one refers to these large-scale operators—whether farmers or managers—many folks, especially those still trying to farm full-time, argued that these people and the current agricultural system in the US put excess pressure on small and midsized farmers. "You've got some big row-crops farmers that are rooting a lot of the little guys out," explained Pete. "It's sad." And it's not just sad, as several people said. It's also broadly harmful. Beyond the environmental consequences of industrial agriculture, there are also negative impacts dealt to rural communities' overall health and vibrancy when many smaller farms are erased in favor of a few massive ones. "There's a strong reason to be deeply concerned when instead of having ten midsized farms producing incomes whose owners spend it in town, you replace that with a large farm whose profits go running off to [big cities]" or end up in the hands of just one local individual or family, said Peter Carstensen, a professor of law emeritus at the University of Wisconsin who was interviewed for a 2019 *Time* article on the topic.[10]

While this situation affects farmers of all kinds in locales across the nation, tobacco production is a particularly illuminating example in Robertson County. This community has long been known as the dark-fired-tobacco capital of the world. This type of tobacco—which is cured by hanging harvested stalks in specialized barns and then building smoldering, smoky fires on the ground to flavor the leaves—was a staple of small and midsized farms up until the late 1990s and early 2000s. Speaking of the prevalence of tobacco in her community, Sarah said, "It was enormous. Everybody had tobacco."[11] Often, farmers grew only a small amount of tobacco—roughly one to ten acres—in patches, explained Denise. She and several others noted that most of the work for a tobacco crop was done with family labor alone. Sometimes, families "swapped work" to help each other out in a pinch.

In 2004, the government price support and production controls for tobacco that had been in place since the Depression era—which largely helped small farmers—were removed.[12] At that point, some farmers jumped headfirst into amplifying tobacco production, leaving farmers who wanted to remain small or midsized out to dry. Brian noted this sudden shift:

> Back in the day, when I was a kid and up until about twenty years ago, every farmer—part-time or otherwise—had a tobacco crop. And the tobacco crop was an important thing because you could net more money per acre off that little spot than most anything else that you did. When the price-supporting structures went away . . . with the tobacco buyout, a lot of farmers opted to sell out their tobacco bases, me included. That opened up the tobacco industry to the tobacco companies to go in and do what they wanted to do unregulated. I don't necessarily agree with this philosophy, but the fewer people you have to deal with to procure your main raw products, generally it's more economically feasible. And that's what they've done.

As echoed in other research, tobacco companies would rather work with a few very large producers than a lot of small ones.[13] The small "patches" that Denise described turned into full-fledged "fields." It's now commonplace for some tobacco farmers to raise hundreds of acres of this crop each year. These farmers—or, perhaps more accurately, managers—often rely on migrant farmworker labor to plant, tend, and harvest crops, which is backbreaking work done in brutal weather conditions. These farmworkers come voluntarily, but their living and working conditions are tough, to say the least. While farmers who transitioned away from growing tobacco did get a small "buyout" from the federal government, the loss of this annual cash crop hurt many smaller-scale farms in the long run. One of their major annual income sources was gone.

People shared that whether in tobacco or other farm types, the growing influence and general expansion of big agriculture has had a pronounced effect on the well-being of small and midsized farmers.

Massive tobacco field that stretches to the cornfield (background, right) and the tree line (background, left). (Photo by author)

Still, large farmers in the Robertson County community are not always viewed in a negative light. While some participants had charged words for big farmers—accusing them of "collecting welfare" via subsidy programs—others viewed them favorably, or at least indifferently.[14] This reaction suggests that frustration with large-scale agriculture is directed more toward the system as a whole and not always individuals. In fact, several people spoke highly of a few big farmers in their area, identifying them as friends and respected community members who participate in local functions and donate to charitable causes. Multiple farm owners interviewed for this project even mentioned renting some of their land to big farmers. Doing so brings in reliable income to help cover property taxes and keep their smaller farms economically viable.

On that general note of nuance when considering large-scale farmers' impact in Robertson County, farmers and nonfarmers alike did, for the most part, prefer the expansion of big farms to farmland loss from development. The loss of smaller-scale farms to big enterprises is far from ideal, they said, but at least land was not being transitioned away from agriculture forever. While "it's not the family farm" as Logan sees it, he believes that "it's still farming. It's not being subdivided with houses built on it. That to me is a good thing." This thinking further signals that many people feel more frustration with the agricultural system and economy in general than with specific individuals.

4

Neither Getting Big nor Getting Out: Imagination, Affection, and Fidelity in Action

The structural challenges facing small and midsized farmers in Robertson County—and across the US—are great. And while many farmers have either succumbed to or purposely embraced increasing real estate development, rapid rural community change, and agricultural consolidation, some farmers are continuing to act as careful stewards of their land. They are neither getting big nor getting out. Instead, they are practicing perseverance. Their hard-earned stewardship virtues help show why.

Virtues are powerful stewardship stimulants because of their deep and enduring nature. Rather than temporary emotions, virtues stick with people and help them determine how to live. By practicing and nurturing stewardship virtues over a long period of time—and observing these virtues in others—a person learns to grapple with and work through difficult situations. People will at times struggle to act on these virtues. This is inevitable. In some cases, external forces—such as having to choose between caring for family or keeping a farm, as described in chapter 3—will prohibit farmers from acting in the best interests of their land, and that's understandable. To be sure, our modern agricultural system itself also throws wrenches into the efforts of those who want to prioritize good care. But as long as they work toward good ends that benefit their own character and serve the people and places around them whenever possible, they are still practicing virtue-based stewardship.

From a philosophical perspective, Wendell Berry's concepts of imagination, affection, and fidelity align with this virtue framework. Perhaps more importantly, they are reflected as virtues on the ground. While the people I interviewed used different words and terms to describe commitments to caring for their land—and while only a handful of participants had heard of or read Berry—nearly every farmer interviewed, and even several nonfarmers, spoke to the presence and power of stewardship virtues. They expressed deep and evolving knowledge of their land, strong and enduring love for their farms, and long-term loyalty and devotion to their places.

Imagination

Imagination means knowing a place intimately. Akin to attunement or familiarity, it's a form of localized knowledge that grounds a person in a place and evolves over time. It also connotes connection and attachment. As a virtue, it is valuable in and of itself and is the foundation on which affection and fidelity stand.[1]

The farmers I talked with spoke often of this notion of imagination. In many instances, their cultivation of this virtue began in childhood. "I feel like I know [my land]," said William, reflecting on his lifelong involvement with his family's farm. "I mean, I've run over it all my life. Probably wasn't paying no attention to it then, but I know where the rocky part is, where water stands longer than it should, stuff like that." Bradley shared similar stories and explained that his children have developed knowledge of and attachment to the farm in the same ways: "I know this four hundred acres like the back of my hand. I ran over it as a kid. My kids ran over it when they were little. I roamed the woods. I know every nook and cranny of it." Other farmers shared the same sentiments, explaining that their connection to the land began when they were young. For Justin, play was an integral part of honing this virtue. When scampering through woods, barns, and fields, he and his sister started to develop an understanding of their family's land. Randy, William, Sarah, Phillip, and Greg also mentioned how romping around the farm as kids helped them foster attachment and attunement to place.

The early development of imagination also stemmed from family members who passed on place-based wisdom, a theme Berry discusses in several of his essays, poems, and novels. Many farmers described becoming closely acquainted with farms through grandparents and parents who served as exemplars of local knowledge. Sarah explained how her father helped her grow her understanding of their place and farming in general. "My dad knew how much I loved to farm. He taught me everything he could teach me. But having grown up here, I was with him all the time. When he was in the fields, I was there every time I could be. . . . I wanted to be everywhere my dad was. I wanted to learn everything I could and be with him and be a part of it." While Sarah herself does not live on a Century Farm, this type of passed-down knowledge was especially common for Century Farm owners. Reading a journal entry from her late husband—whose family has owned their farm since the 1880s—Marie highlighted how "the secrets of this place" were passed down from parents to children. Other farmers also described how having multigenerational connections to land helped them know their farms intimately.

James Rebanks—an English farmer and author whose writing and work is influenced by Berry—also speaks to nurturing imagination for place as a child. Speaking of early experiences around livestock with his grandfather, Rebanks writes, "Time seemed to slow down around my grandad. He believed in watching carefully and taking time with his animals. He would simply gaze at his cows or sheep for what felt like ages, leaning over a gate. . . . He believed that a good farmer was patient and used his, or her, eyes and ears, and nose and touch. . . . He called me his 'squire,' which I never understood until much later—I was his project, his apprentice."[2] Rebanks may not have realized it then, but his grandfather was acting as a teacher and model of imagination. He was showing the younger Rebanks how to pay attention and put traditional, place-based knowledge to good use. These lessons stuck with the now-famous shepherd. The foundation of his farming is imagination, rooted in childhood experience.

Not everyone I interviewed, however, grew up on the farms they currently care for. Their imagination didn't develop until later in life. Still, they have practiced this virtue diligently. James is a good example. He moved to his small farm in northeastern Robertson County roughly thirty years ago. In that time, he has learned a lot about his land, down to very specific details. For example, James described a seed-sowing and mowing method that he has refined over years of observation to fit his primary hay field. "I'm not fighting the field," he said. "I'm taking what it's giving me. It's enough for my cows." His knowledge of the field and what it can give, the hay's growth and reestablishment patterns, and the needs of his cattle reveal the practical benefits of careful attention and imagination. James's comments also illustrate a welcome understanding of what is "enough" for himself, his livestock, and his farm, a humble concept frequently forsaken in favor of pursuing abundance at all costs.[3]

Pete also serves as a helpful example, though for a different reason. He grew up on another nearby farm, where he learned a lot about agriculture from his parents. (Pete's brother still works and lives on their family's farm.) But he moved to his current farm—where he and his family milk a small herd of cattle, an increasing rarity in dairy—about fifteen years ago. After I heard Pete talk at length about his Jersey cows, his fields, and his woods—he was speaking with me over the phone while waiting for a tire to be repaired at a local shop, so to my delight, he was in no hurry with his remarks—it became obvious that his imagination is fine-tuned. He spoke in clear detail about the land and the animals that call it home, so much so that I could almost envision the place even though I'd never seen it. But when I made a comment about how well he seemed to know his land, Pete laughed and replied, "I'd like to know it more." His humble, offhand comment shows that he—like others interviewed for this book—sees his knowledge as incomplete. Or in virtue theory language, Pete sees imagination as something that he must continue to practice. Just as important, it's something he *wants* to do. Brian offered a similar statement, saying that while he is highly familiar with his farm and has spent decades on it, he is "surprised by it all the time, by new things that happen or develop." Noting changing soil conditions, shifting wildlife movements, and the continuing evolution of his fruit trees, he mentioned that "even though it's not a surprise a minute, there's always something new." Again, these comments stress the need to continuously nurture imagination and emphasize why it is more than stagnant knowledge.

Comments from land-conservation professionals also highlight the imagination that some small and midsized farmers in Robertson County have honed. Melissa explained that someone from her organization always visits a farm before the organization commits to working on a specific conservation project. Usually, the farmers who express interest in protecting their land from development via conservation easements—and especially those who follow through with this permanent decision—have "owned their land for a long time, or it's a piece of family land": "So people really know their land well. They can take us to their favorite scenic spot. They can show us the springs. They can avoid the holes in the field. It's really important to be out there and see the property with those people so they can show us those places."

Paralleling these comments, Nathan offered his own take. When he visits a farm that is protected from development to ensure conservation agreements are being upheld, he carries a binder filled with maps, photographs, soil reports, habitat information, written narratives, and more. He spends hours before each visit reviewing these records, which can add up to well over one hundred pages of materials. Even though he is armed with an ample amount of scientific and spatial knowledge about any given farm, he said, "[The farmers] still know more about [the land] than I do, especially some of the farmers who have been on their land for forty, fifty, sixty years. They can tell you just about everything that's out there, and they can tell you just about everything that's ever happened in the time that they've been working it or owning it." Many times, these people also have an understanding of the farm from before their time. This localized natural and cultural history is passed down through stories from older family members and neighbors.

In addition to enhancing on-the-ground knowledge and understanding of a farm, imagination also breeds attachment. Most of the farmers interviewed described the land as an extension of themselves, with one person, Marie, even saying, "For me, it's just like a part of your body. It's like your other arm or something. I can't imagine it not being part of us. I just can't even think of what it would be like not to have this [place]." This feeling of physical and spiritual oneness with place is like Berry's description of his own imagination, described in chapter 1. He explicitly says that he grew to be aware of his farm as if it were part of his own body.[4] The phrase "This land is a part of me"—or a close variation of it—was used frequently by farmers. Brian shared this sentiment:

> [This farm] is a part of me. Everything that's here on this farm, with the exception of three buildings and some trees that are scattered around, I had my hands on it or was involved in making it happen. So when I look around and out the back door, and I see the farm shop and the hay barns and the fences, the orchard, all of those things, it's almost like a baby. It has been my thing to do. Everything—and not every waking minute—but there has been lots and lots and lots of time and effort, from plotting and planning and

figuring out how we're going to make this happen to figuring out how we're going to fix it after it screws up and making it all work through that.

Brian's close connection to all parts of his family's farm, his continued evaluation of the place's condition, and his comparison of the farm to a baby, which suggests both tenderness and fierce attachment, reveal imagination in action. His remarks also indicate how affection can sprout from this foundational virtue.

Affection

As a stewardship virtue, affection is a deep-seated disposition of a person, one that endures through adversity. In this way, it is far more than a passing feeling or emotion. Like imagination, it must be continually practiced. Affection must also be directed toward good ends, both for the person who nurtures it and for the land and community—human and nonhuman—that affection is directed toward. It stimulates care.

Many of the people I interviewed indicated through their comments that they actively practice virtuous affection for their farms. But rather than use the term "affection," they favored "love," a more common word. Despite the slight difference in vocabulary, they described love for their farms in ways that align with the key characteristics of Berry's affection.

As with imagination, several farmers' affection for the land was shaped by family members. As Justin reflected on how his parents taught him and his sister to cultivate this virtue, he said, "They really inculcated in all of us this respect and love for the land." But Justin did not fully understand his affection for the farm until he moved away from Robertson County to attend college. At some point, he recognized that "it's very strange and unique and special to have a connection to a specific plot of land": "I realized how lucky we are to have it." Like Justin, many others learned to love the land from older family members and, in some cases, community members.

Beyond observing and learning from exemplars, farmers described refining their affection through action. When Greg was pressed about his professed love for his farm and asked how that love developed, he laughed and said, "You just learn from doing," echoing the habituation at the heart of cultivating virtues like affection. He went on to describe the work he has done in hay fields and with cattle, sometimes with family and sometimes alone. These moments sharpened his affection. Robert expressed these thoughts too. When we spoke on the phone, he had just come inside from a long day of picking pumpkins, stoking fires in the tobacco barn, and cutting hay.[5] These larger jobs aside, he had also spent much of the day handling dozens of smaller tasks. He was worn out, exhausted. As with Greg, I pressed Robert when he said he loves his land, asking for specifics. He went on to explain that his affection is anchored and fine-tuned in the work that he and his family have done for over a century on the land. "I get awful tired and grumble and get disgusted sometimes," he said, alluding to elements of agricultural adversity and his aching back. But he keeps at this work out of love. His words echo some Wendell Berry shares in "Notes from an Absence and a Return." Here, Berry describes returning to his farm after an extended time away: "As never before I'm impressed with the dependence of a human place, such as a farm, on human love. After our seven months' absence our place shows clearly its dependence, not so much on the conscious large acts such as a [person] might do out of duty, but on the hundreds of trivial acts that a [person] who loves it does every day, without premeditation, in the course of doing other things."[6] These stories from Greg, Robert, and Berry show that affection often flourishes in the day-to-day mundane.

Others told similar stories. Sarah spoke about how her love for the farm motivates her to endure tough weather and work conditions. In the winter, she wakes up early each morning to feed hay to her cattle. (Sarah rolls her hay bales out so that one area isn't mucked up during feeding. This feeding method also helps spread out her cows' manure, she explained, which benefits the health of her soils. Because she feeds in this way, she puts hay out daily instead of feeding multiple bales in rings and letting cattle eat until the hay is gone, as many farmers do.) As a part-time farmer, Sarah must finish this work before she begins her full-time job off the farm. Many mornings are "rough and nasty and cold and whatever," she said. Sometimes, these conditions get the best of her, like when she tripped on a root last year, cutting her head during the tumble and damaging her knee to the point of needing a replacement. The injury put her out of commission for a while, so her daughter took over feeding duties. But still, Sarah emphasized, "When I go out and feed every morning in the winter—*every* morning—I love it." It is, she explained, her affection for the place—for the land, the cattle, the wildlife, the people in the past and present who have called the farm home—that keeps her going in tough times.

Bradley also shared a compelling story about entrenched affection helping him persevere in his work as a farmer. Referencing a line from the Future Farmers of America (FFA) Creed that he learned in high school—"I know the joys and discomforts of agricultural life and hold an inborn fondness for those associations which, even in hours of discouragement, I cannot deny"—Bradley told a story about caring for pigs.[7]

> When I was about nineteen, I was going to college and farming, and I had about thirty sows of my own. But the only time I could work with them—because we were so busy—was when it rained. [*If it wasn't raining, Bradley would be working in crop fields to take*

advantage of dry weather.] And these were outdoor hogs. So when it rained, I would put on my boots and go out there in a steady rain and do fencing, move feeders, do whatever. One day, I was trying to move a feeder, and you know, a pig lot is packed like concrete with about a half-inch of slick mud on top. I was trying to move this big feeder, and I went to pull on it, and my feet slipped out—I sat right down on my butt in a mud hole. And the first thing that come to my mind when I hit the ground was, "God, I love this."

On the surface, Bradley's story is comical. We both laughed when he shared it during the interview. In fact, I probably laughed too much. But beyond its humor, the story also reveals how affection can encourage persistent care during difficulty—such as being soaked to the bone and covered in pig shit but continuing to find purpose and joy in the work.

While the preceding comments and stories help illustrate the presence and power of virtuous affection, the most compelling remarks came in response to simple questions. During each interview with a farmer—usually after they had spoken about the challenges they face—I asked, "Why do you continue to farm? Why do you hang on to this place?" As noted in chapter 1, these questions are rooted in one that Wendell Berry poses in his essay "Conservationist and Agrarian."[8] While a few people mentioned economic, familial, or religious reasons, most farmers—as Berry anticipated—said that love was their primary motivation.

Addressing the aforementioned questions for himself and speculating on why he thinks other small farmers continue to tend their land, Logan said, "I think it's a love of the land and a love of the way of life. Like myself, it's the way we were raised. That love of the land, that deep-rooted connection to the earth that drives most of us that are part-time farmers or whatever." Adding to her earlier comments, Sarah's response was also rooted in affection. "Because I love it," she said in answer to my question. "I just love it. I love the land, and I

One participant's farm in eastern Robertson County. Note the birds in flight, cross fences, resting cattle at far right horizon, lush summer grasses, and old tobacco barn. This farm is healthy and productive, yet it's still endangered.
(Photo by author)

have such an attachment to it. It is not for the money, for sure. It's not for the money because my expenses almost always outweigh what I'm making. Sometimes I break even. When I sell cows for grass-fed beef, I do a little better. But no, it's because I love it. It has been my life, my entire life. And I love it." In fewer words, Carl also claimed affection as his primary stewardship motivation: "I deeply love this land. I know in my heart that this is my calling."

On the basis of frequent interactions with small and midsized farmers, even nonfarming participants felt that affection was a driving force in motivating stewardship. Benjamin spoke from his experiences working with farmers on soil-health matters, saying that he thinks love is a primary reason many small and midsized farmers persevere and practice conservation despite development pressure and agricultural consolidation. Frances did too. Responding to a question about what motivates people to keep farming in the face of adversity, she said, "Well, I think it all goes back to why they do it in the first place. And I think it's that connectedness and love of the land and love of growing things. It's a special connection to the land. I think a lot of them could quit and do something else, and maybe they'd make more money doing it. But that's what they love. I think it's a deep-seated emotional connection." While the deep-seated connection she references is more akin to virtue than emotion, Frances's point stands. Affection is key.

John's response to questions regarding stewardship motivations and perseverance is worth quoting at length. He grounded his thoughts in his decades-long work as a town manager in a small Robertson County community, which has led to close friendships with farmers.

> It's a good question: Why do they do it? Why go under the stress and the pressure and worry about the rain and all the insects and all the challenges that come with farming? If you're not making much money at it, why do it? I think it has to be your connection to the land and what's been ingrained in you by other generations. I think most people here were brought up being told the value of hard work and having your hands in the dirt and working your farm, and I think that's just something that just doesn't come out of people. I think that's how they see life. A lot of those people measure success not by the dollars in their bank account but by whether they enjoy what they're doing or not. . . . They bitch and moan about the prices they're getting or the drought or all the other things, but at the end of the day, they farm because they love it.

Addressing many elements of the virtue theory framework—having "ingrained" dispositions, learning from exemplars and personal practice, grappling with adversity—John highlights on-the-ground implications of virtuous affection.

While affection is actively practiced and observed by most of the farmers I interviewed, one couple, Larry and Carol, initially seemed to trivialize it. At one point in our interview, they laughed and said, "If you don't love something, you don't spend money on it!" This thinking appears shallow at first, but their point is valid. A recent article analyzing farmer livelihoods across the US found that "though many farm households have incomes comparable to the US average, much of this revenue comes from off-farm activities. On-farm revenues are often eclipsed by on-farm costs, meaning that in many regions of the US, farm operators *pay* to engage in the labor- and time-intensive act of operating a farm."[9] Affection can require investment and financial commitment.

Yet others' affection was more intense. Like Brian earlier, Phillip compared his affection for the land, which has been in his family for more than a century, to the love someone has for a child. In our interview, he told me to imagine that I had children and then asked how much money it would take for me to sell them. I was confused and, to be honest, startled. "Whatever you sell your kid for, I'll sell my land," he continued. The implication? Because of his forceful love for the land, no price could ever be high enough for Phillip to sell his farm. It would be a violation of authentic love, just as it would be for a parent to sell a child. For many farmers, love is the primary reason that they continue to act as "stickers" and stewards. Their fidelity stands on authentic affection.

Fidelity

When anchored in imagination and affection, fidelity encourages long-lasting care for a place, combining elements of loyalty, responsibility, devotion, and duty. A few sporadic acts of stewardship do not denote fidelity, but a sustained commitment to the well-being of a place and its community does. And as with imagination and affection, fidelity should endure through adversity. It is in all respects a deeply rooted disposition, a hard-earned character trait that is necessary for loving care. This is why Berry uses marriage as an example to illustrate fidelity to the land. Every union faces tough times, but when a relationship is rooted in understanding and love, commitment and care become welcome responsibilities that benefit both members and spark joy, wholeness, and fulfillment.

This joyous obligation to care for a place even through difficulty was mentioned by many farmers. Carl described efforts to stave off bankruptcy decades ago, dirty dealings by agribusiness companies, struggles with farm-related physical injuries, recent economic challenges from the COVID-19 pandemic, the cost of providing care for his ailing parents, and more, explaining, "I've spent my whole life fighting to keep this farm together, literally fighting to keep it together. And it has been just one thing after another after another after another. You get one thing put to bed, and it's something else. . . . Honestly, I don't really know how we got through it all." Carl said that he takes his "responsibility" to care for the farm "very, very seriously," and his example shows the power of fidelity in sustaining stewardship.

Bradley shared a similar story. When he was around twenty years old, he and his dad lost their farm. "We went bankrupt in the '80s, and we lost everything," he said, referencing the infamous Farm Crisis that affected so many farmers in that era. But in an act of grace, the local bank offered a second chance. Bradley explained that the bank gave his family a year to come up with a down payment. If they could pull together this money, the bank would let them buy back the farm. They would have to start the mortgage payments over again, but the land would stay in their family instead of being auctioned off. Bradley and his family scrounged and saved, working incredibly hard to save their farm. They were able to make that down payment. Reflecting on what it took to hang on to their land and "all the blood, sweat, and tears" spent in the process, Bradley proudly said, "We've bought and paid for [this farm] *twice*."

Others discussed the sacrifices they have made to continue tending their farms too. Economic challenges loomed large. For this reason, a large percentage of the farmers I interviewed either currently work or have worked full-time jobs off the farm to generate income. At least fifteen people described relying on off-farm income to support their farms, a percentage comparable to national statistics that describe farmers and off-farm income. The constant work that this lifestyle requires—farming before and after a full day's work, as well as farming on the weekends—leaves some folks feeling exhausted, even if satisfied and fulfilled. But the extra work and income are necessary if many of these people hope to keep their farms, especially since they are not farming on massive scales that generate more profits. In a representative comment, Phillip said, "I feel like you got to work off the farm to keep what you got." Similarly, Logan explained, "I work at a public job so we can keep farming. I'm sustaining my way of life. . . . That's what I'm doing." Many others echoed these thoughts and actions.

Given the various economic challenges that smaller-scale farmers face in holding on to their farms—challenges that force them to fight through all kinds of tough situations, buy the same land twice, and finance their farms with off-farm labor—a reasonable question looms. With such a hot market for residential and commercial development in Robertson County, and with large-scale farmer-managers always hungry to expand their operations, why don't the farmers featured here act in a way consistent with market forces? Why stay rather than sell?

Despite the high prices they could get for their farms—prices that, as noted earlier, can quickly exceed $20,000 an acre—many people expressed that they have no interest in selling their land, even if it would make their lives easier. Sarah, for example, mentioned that her farm has several characteristics that make it financially valuable. In addition to its flat, fertile, and open pastures, which make the land highly developable, the farm is just "a mile from a junior high and an elementary school. It's very valuable property. But," she continued, "there's nothing that could make me sell this farm." Likewise, William—who, as he mentioned earlier, lives near Interstate 65 and has watched homes be built right up next to his property boundary—said, "It has never crossed my mind to sell one inch." Frank put forward the same sentiments. He explained a recent conversation with a man who kept asking to buy some of his family's farm to build new homes: "I had one guy here about a month ago who said, 'Just sell me any part.' I said, 'Naw, we're not going to sell none of it. . . .' He said, 'I tell you what I'll do, Frank. Any place you face the highway, I'll give you $40,000 an acre. I'll write you a check right now.' I said, 'I wouldn't sell it to nobody at all.' " Frank easily had more than a million dollars sitting in his hand from this offer, but he turned it down without hesitation. For Sarah, William, and Frank, their fidelity to place outweighs putting more money in their pockets.

Other farmers echoed this loyalty to the land. Denise mentioned that if there were a medical or other financial emergency, it is possible that her family might need to sell some of the farm. As with Carl's earlier story about selling part of his family's farm to provide care for his sick parents, Denise's position here is certainly understandable. But she then said, "I have no plans to sell the land at all. I've actually even tried to consider if there's a way to buy a farm that's being sold, just because I hate to see it subdivided." Continuing, Denise spoke about the duty she feels to care for the farm. "The easy way out would have been to quit a long time ago. I have a lot of road frontage here, and I could have divided it and been gone and been financially well-off. But we don't want that path. We're willing to scrounge each year and keep working at it." In a more succinct manner, Randy said the same: "If somebody offered me $20,000 an acre [for this land], I'd just tell them to go to hell." Using tamer verbiage, other farmers shared comparable outlooks.

In several cases, farmers described trying to pass this disposition down to their children. They aim to be role models, they said, and hope that their children cultivate a similar commitment to the land. Perhaps unsurprisingly, Berry speaks to the importance of "training" the next generation of caretakers and farmers in many of his writings. In an interview with Amanda Petrusich of the *New Yorker*, he makes his belief in the power of exemplars clear: "When farmers are taught, starting in childhood, by parents and grandparents and neighbors, their education comes 'naturally,' and at little cost to the land. A good farmer is one who brings competent knowledge, work wisdom, and a locally adapted agrarian culture to a particular farm that has been lovingly studied and learned over a number of years. We are not talking here about 'job training' but rather about the lifelong education of an artist, the wisdom that comes from unceasing attention and practice."[10]

Some of the people I talked to shared that they can already see this imagination- and affection-fueled fidelity in their children—and, in some cases, even their grandchildren—which makes them feel confident about the continued stewardship of their respective farms far into the future. But even those whose children are less interested in agriculture or who want to move to a new area mentioned hoping to find and support future caretakers who will love and respect the land. Pete, for example, acknowledged that his children may decide not to take over the family farm. While he would love for them to continue farming, he said he will respect whatever decision they make. He understands their uncertainty given the difficulties inherent in smaller-scale agriculture, hardships that his children have witnessed firsthand. If they do decide to pursue another path, Pete said, "I would like to help somebody [else] and let them take [the farm] and run with it when I get tired, if that happens. . . . I'd like to see it go forward." Similarly, Janice and Randy, as well as Kyle and Phillip, explicitly mentioned dreaming of finding the next generation of nurturing stewards for their land, even if they are not part of the family. They aren't opposed to selling their land if the right situation arises. They just want to ensure the place ends up in the hands of people who will care for it as they have.

The preceding stories, thoughts, and experiences are all helpful for illustrating fidelity to place. But a few farmers' actions stand out as ultimate expressions of devotion. Six farm families partnered with a nonprofit land trust to place conservation easements on their land, ensuring that their land will remain open and undeveloped in perpetuity. These legal agreements place permanent restrictions on properties, effectively meaning that the conserved land will remain farmland or open space forever, no matter who owns it. These families are willing to make a significant and permanent financial sacrifice to care for the land they love.

When farmers talked about their motivations for making this lasting decision—and simultaneously lowering the market value of their land, given self-imposed development restrictions—they spoke indirectly about the virtues of imagination, affection, and fidelity. While a few mentioned that they received some tax benefits for protecting their land, most spoke about their intimate connection to their farms as their reason for conserving them. "When you've grown up with it and you see how important it is and how beautiful and wonderful it is," mentioned Sarah, who partnered with a land trust to conserve her land years ago, "then there's not a price on it, which brought me to [a land trust]." Combined with the sentiments she expressed earlier about imagination and affection, fidelity clearly inspired Sarah to make this decision. Melissa, who has worked with multiple farm families in Robertson County to help them protect their farms from being carved up and developed, spoke further about fidelity's power. She noted that while there are exceptions, most people who decide to place a conservation easement on their farm have an "underlying care and love for the land" that motivates their fidelity-based actions.[11] Beyond those who have already permanently conserved their farms, several other participants, including elected leaders and county officials, explicitly mentioned supporting land conservation. Some said they hope to protect their own land soon, if they can afford it.

Through a variety of comments, stories, and actions, the people I interviewed show how fidelity functions on the ground. While these farmers acknowledged that easier, more lucrative paths exist, most remain committed to continued care. Sustained stewardship means more work. It means fighting off economic challenges. It means enduring the pain of watching neighbors' land being bulldozed and developed or incorporated into massive farms. Still, some farmers' fidelity encourages and enables them to persevere.

Other Stewardship Motivations

Although most of the farmers featured here lifted up the combined virtues of imagination, affection, and fidelity as the primary reasons for continuing to farm and tend their land as their community changes because of development and large-scale agriculture, a few also cited other motivations. Of these alternative stewardship stimuli, economics, religion, worldviews, and family stand out.

While most of the people I interviewed stressed that the financial prospects of smaller-scale farming are somewhat dim at the moment, a few mentioned earning income as a key reason for continuing to farm. Some—especially those who farm full-time and rely entirely on agricultural income—held economic motivations alongside virtue-based motivations. Kyle stated, "Farming ain't all about profit. It ain't. If that's all you're thinking about, then you're losing what all this is about. It ain't all profit." But he and others also explained that earning money is important for securing long-term stability, keeping the land, and providing for family. Henry said that earning a profit from selling crops or livestock "gives you a sense of at least [the land] is paying its way, things such as that." If nothing else, folks who also earn off-farm income hope that they can make enough money from their farms to cover their expenses and pay their property taxes. Maintaining or enhancing a farm's economic viability, then, is a key factor in securing its future and should be thought of as a conservation strategy.

Spirituality and religious beliefs—particularly those anchored in Christianity, given the religious culture of the area—are also important for some farmers.[12] For Janice, her faith pushes her and her husband to care for the land: "God told us to be stewards of this creation. We're not owners of it. We're stewards. And good stewardship, to me, is the same as a sound moral life. It is one and the same, where I am not the center of the universe and my decisions should be based on what is best, what God would have me do for others. That should be my source of decision-making." William also expressed these thoughts, saying, "I think [my faith] plays a part in the way I care for this place. I think the Lord gives you that [land], and he wants you to take care of it. I think it's your responsibility to take care of it."

Bradley, too, views his Christian faith as inspiring care of creation. After lamenting how, in his eyes, people have stopped seeing land as a gift from God—which has led to land being treated as "just a commodity to be traded," especially "for these people who [build] houses" on farmland—Bradley noted, "I think we're supposed to be good stewards and take care of what the Lord has given us." Bradley's comment aligns closely with some of Aldo Leopold's thinking. While Leopold's is not exclusively a religious perspective, he writes in the foreword to *A Sand County Almanac*, "We abuse land because we regard it as a commodity belonging to us. When we see land as a community to which we belong, we may begin to use it with love and respect." Place is not to be exploited. It is to be treasured and tended carefully.[13]

Less overtly Christian thoughts were shared by farmers too. Some who do not identify as religious shared the specific spiritual values of connection, awe, respect, gratitude, and soul nourishment—as well as the profound spiritual impact of watching things grow and die—as reasons for persevering on the land.

Family legacy—in both the past, present, and future—is seen as influential too. For farms that are multigenerational, several farmers described feeling a responsibility to the family members who came before them and those who will come after. Selling family land for development or industrial agriculture would be, for some folks, an affront to the hard work of past generations and a severed connection for future farmers. Capturing these thoughts concisely, Bradley explained, "My grandparents worked so hard to get this land. We've worked hard to get this, and I worked and struggled to get this. And you hate to see all that just blown up." Other farmers also spoke to the meaning of family legacy and mentioned how they have tried to instill this concept into their own children.

Beyond economics, faith, and family, some farmers referenced how a general worldview shapes their stewardship. Several mentioned decentering themselves when thinking of long-term care for their farms. They noted respect for future generations, the health of the land and earth, the needs of nonhuman animals, and the well-being of their community. Expressing a mantra that prioritizes selflessness and the health of the land, the planet, and, in turn, people, James explained, "The general statement that my beliefs come from is that mankind needs to live with the earth, not at the expense of the earth. And it's that one word, four letters: *with*. We need to live *with* the earth, not at the expense of the earth. I don't own this land. I just live here. Somebody else will have it after I'm gone and somebody after that. . . . Just live with the earth. When you do everything—well, not everything, but a lot of things for your own benefit—what about the earth?" James's outlook parallels the thinking and beliefs of many well-known environmental thinkers, including Berry.[14]

It is important to recognize that these other motivations for stewarding farmland are not exclusive of the virtues of imagination, affection, and fidelity. In fact, such elements as religion and spirituality, family, and land-ethic-like worldviews are often sources and supports for these virtues. Economic motivators—when viewed in a nonexploitive and sustainable manner—can also help facilitate a person's ability to nurture imagination, affection, and fidelity by providing financial stability and security. Farming is, after all, partly a job, and people who farm need sustainable and fair income. These various motivations can blend with stewardship virtues to reinforce care.

Small and midsized farmers in Robertson County face a flurry of challenges. Residential and commercial development are remaking the landscapes that they have long known, replacing many fields and forests with subdivisions, large-lot housing tracts, and warehouses that reflect rural gentrification. This mass erasure of agricultural land has caused real emotional pain for many longtime community members, not to mention cultural, environmental, agricultural, and economic trauma. At the same time, agricultural consolidation is incorporating much of the county's remaining farmland into massive operations. Participants see smaller-scale farmers as systemically disadvantaged, and they worry about the future of farming in their community and across the country. Coupled with the everyday difficulties of farming and farm life, small and midsized farmers who want to keep caring for their land and aspiring farmers who aim to lead an agrarian life have been dealt a hard row to hoe.

Still, many of the people I interviewed for this book show a serious commitment to stewardship. They recognize that their path is not an easy one, that selling the farm—either to a large-scale farmer-manager or to real estate developers—would lead to windfall profits, and that their fighting and clawing and scrounging could stop. Their struggles could be wiped away with a few "for sale" signs. But many of these farmers have chosen—and continue to choose—to remain on their farms, to be "stickers" and stewards. For such farmers, "the effort of conservation has ceased to be a separate activity and has come to be at one with their ways of making their living," Berry writes. "They have not achieved perfection, of course, but they have achieved a kind of unity of vision and work."[15] Their imagination of, affection for, and fidelity to their places on earth—and the various communities who also call these places home—have led them toward thoughtful, loving, and lasting care. In their example, we see virtues in action.

We can learn from these farmers. Although they are imperfect and readily admit their own shortcomings, they can be understood as everyday exemplars of stewardship. Their example can, of course, be instructive to other farmers in other places, perhaps encouraging them to practice stewardship virtues in the face of adversity too. But their example can also transcend agriculture and enter the field of environmentalism. While farmers offered mixed responses when asked if they identify as environmentalists—for those who did see themselves in this way, they almost always localized their environmentalism, focusing on how they care for their specific farms—their words and actions carry important lessons for the environmental community and the world.

These small farmers have connected with and become responsible for the well-being of the land, even when doing so is difficult. Intimate relationships with place and their continual honing of imagination, affection, and fidelity push them toward providing care. People have denied themselves economic gain, ease, and reduced labor in favor of the hard work of stewardship. They have chosen this labor of love.

In an era of unprecedented environmental and ecological challenges, more people must embrace actions that are difficult, that are personal, that prioritize care over profit alone. This is not to say that everyone should flock to small farms, though we could collectively use a steep increase in the number of well-trained and committed small and midsized farmers rather than the continuing dominance of industrial operations and corporations. But it does suggest that people should foster and uphold place connections and commitments in other ways and locales. Doing so can help develop or strengthen virtues that make environmentalism active, resilient, and ingrained.

The stewardship virtues embodied by some small and midsized farmers in Robertson County lead to real, tangible good. And with intentional work, they can be emulated. The people I interviewed show how imagination, affection, and fidelity move from philosophy to the farm. There, these virtues have grown. If we choose to look, listen, and learn, the yield may yet be bountiful.

5

Systemic Struggles:
Racism, Sprawl, and Consolidation

Walter called me from his car on a Sunday evening in early August. We had exchanged several emails over the month prior, trying to find a time to speak. A full-time teacher in the Maury County public school system, Walter had been busy preparing to start a new school year, one that—because of the COVID-19 pandemic—promised to be full of challenges. The next day, he and his students would enter a situation that, as he said, "is probably unique in the whole history of going back to school."[1]

To prepare for the first day of class—and to refresh his thoughts, memories, and feelings for our conversation—Walter spent that Sunday afternoon on his family's farm in rural Maury County. He needed to ground himself, he said, and spending time on his family's land helps him rekindle a sense of connection and determination. "My mother and brother still live out there," Walter explained as he sat in his driveway back in town, talking to me on speakerphone. "My uncle still lives out there. But the house where my oldest uncle lived, nobody's there. I went out there today and just sat on the porch. That was somewhere where generations of people used to sit out on the porch. And I went out there and just sat, just thinking and reminiscing about everything." Explaining that he halfway expected his deceased uncle or grandmother to walk out of the house and join him, Walter noted the power, peace, and purpose he feels when spending time on the farm. Now in his fifties, Walter has felt this sense of belonging for most of his life.

With a decrease of nearly fourteen thousand acres of agricultural land between 2002 and 2017 and many remaining farms consolidating into larger, more industrial operations—the same dynamics confronting rural communities an hour's drive north in Robertson County—fewer folks in Maury County are sitting on rural front porches, looking out and reflecting on the past, present, and future of their farms. And not many of those remaining farmers look like Walter. Demographic data from the Census of Agriculture show that, as of 2017, only sixty-one Black farmers are listed as "producers" in Maury County.[2] To be sure, these sixty-one Black farmers are far more than the eleven documented in Robertson County. In fact, Maury County has the third-highest population of Black farmers among Middle Tennessee's thirty-eight counties. But they are a fraction (2.4 percent) of the 2,592 total producers listed for the county. In addition to the everyday struggles of farm life and the aforementioned issues of development-induced land loss and agricultural consolidation, Black farmers here have grappled with race-based challenges. These obstacles, which manifest themselves in both individualized and systemic forms, have made agricultural survival difficult for farmers of color across the US.

Yet despite the difficulty inherent in continuing to farm and care for the land, some Black farmers in this community persist in place. And while they cite family legacy, spiritual calling, and economic opportunity as some of their reasons for perseverance, the people I interviewed for this book also spoke about the virtues of imagination, affection, and fidelity as motivators of their sustained stewardship. Samuel, who lives and farms in northwestern Maury County, made his commitment to the land explicit. Just like his father before him, Samuel said, "I hope to die here on my farm. . . . I ain't no way soon ready to go, but if I do, I hope I leave here on my farm. We've got a community cemetery down here," he continued, "and my wife has already made plans to put our tombstone down there because that's where we want to be buried. When I get through farming and stuff, that's where I want to go. I love my farm." In life and in death, Samuel's fidelity, rooted in affection, will endure.

Using the same ethnographic methods employed in Robertson County alongside supplementary historical research, this chapter explores several of the challenges that smaller-scale Black farmers in Maury County face. Then, chapter 6 highlights their motivations for continuing to act as careful, committed stewards of the land. In the face of adversity, the farmers featured here have nurtured enduring relationships with place. Grounded in understanding, love, and loyalty, these people-place relationships have lasted, in some cases, for multiple generations. Through these examples, the power of stewardship virtues is shown again.

Why Maury County?

Engaging with and learning from Black farmers was a foundational goal of this research for a few reasons. First, racial diversity is severely lacking in American agriculture. Uplifting the voices of nonwhite farmers thus becomes an issue of justice and representation, as well as a potential means for inspiration. Experts assert that hearing stories of successful Black farmers may encourage more young people of color to explore opportunities in agriculture, even in spite of the painful history of farming for African Americans.[3] Second, given this painful history and the extra burdens placed on Black farmers simply because of their race, discussing stewardship motivations with Black farmers can help better illustrate the roles of imagination, affection, and fidelity. If these virtues have inspired participants to continue stewarding their land in the face of such difficulty—as chapter 6 shows that they have—then their power and potential is even further demonstrated.

Unable to connect with farmers of color in Robertson County, I sought out other communities in Middle Tennessee, hoping to find a larger population of Black farmers who might be willing to speak about their relationships with the land. When digging through data from the most recent Census of Agriculture, I learned that Maury County was home to more Black farmers than almost every other county in Middle Tennessee, behind only Giles (ninety-five) and Rutherford (sixty-five) Counties. Given my familiarity with Maury County—my family's small farm is about ten miles east of the county line—I focused my efforts there.

As in Robertson County, I initially struggled to connect with Black farmers. Local extension agents, farm-store workers, and agricultural service providers kindly offered assistance, but the contact information they shared was unfruitful. Most phone numbers, for instance, either had been disconnected or did not offer the opportunity to leave a message if no one answered. The COVID-19 pandemic complicated outreach efforts too. On three occasions early on, I was able to make contact with Black farmers, but our conversations were limited. Two people cited their busy schedules and explained that they did not have time for a full-length interview. In the other case, I emailed with a Black farmer's adult son, who no longer lives in the area. The father—who has farmed in rural Maury County for decades—would rather not stir painful memories of racial discrimination and agriculture, the son told me. Though supportive of the project, the younger man declined to participate on his father's behalf.

Through continued outreach efforts, I connected with two people in the community who became helpful gatekeepers. Without their help, my efforts may have floundered. One person, who leads an African American history organization in Maury County, put me in touch with three Black farmers after learning more about my research. Another person, who is well-known in the local farming community, introduced me to two more Black farmers. These five people care for farms that are classified as small or midsized—the smallest farm covers 54 acres, while the largest contains roughly 240 acres—so they are keenly aware of the challenges facing smaller-scale farmers.[4]

Despite sustained attempts to reach more people, I was only able to interview these five Black farmers. In addition to the logistical challenges of connecting with people, my on-the-surface status as an unfamiliar white person affiliated with an elite northern institution may have understandably made people hesitant to participate. There are precedents for small sample sizes when conducting qualitative research with Black farmers, but the modest number of participants remains a limitation of this work.[5]

Beyond the ability to better connect with Black farmers, even if in a limited capacity, I settled on Maury County as a field site because of its similarities with Robertson County. Comparable community characteristics are important for the sake of research consistency. Both Robertson and Maury Counties are close to sprawling urban centers. Both have suburban bedroom towns, a centrally located county seat, and distinct rural communities. Both have a deep farming history. And perhaps most importantly for my research purposes, these two counties are both experiencing simultaneous farmland loss and agricultural consolidation.

Data from the Census of Agriculture show that between 2002 and 2017, Maury County lost 13,654 acres of farmland, a decrease of nearly 6 percent. While this figure is much smaller than Robertson County's decline of 41,245 acres during the same period, it is still substantial, equating to a reduction of over 900 acres of agricultural acreage each year.

Expanding the observation window is important to understanding Maury County's farmland conversion. It is especially helpful to look at statistics from the 1980s and 1990s. Between 1987 and 1997, development increased significantly in Maury County, and the county's farmland acreage decreased by over fourteen thousand acres. Much of the new residential and commercial development was related to an uptick in industry.

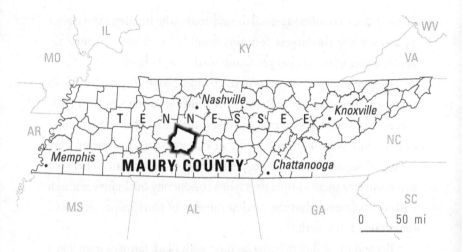

Maury County's location within Tennessee. (Map created by Luke D. Iverson; data sources: US Census Bureau TIGER, Natural Earth, National Land Cover Database 2016, Esri Living Atlas)

In 1990, General Motors (under the Saturn brand) opened an automobile-manufacturing plant in Spring Hill, a formerly rural community near the northeastern corner of Maury County. With this massive factory and associated industries, the town's population exploded. Census data shows that Spring Hill's population grew from 989 residents in 1980 to 7,715 residents in 2000, an increase of 680 percent. For the sake of comparison, Tennessee's population as a whole increased by roughly 19.6 percent over the same period.

With Spring Hill's proximity to Nashville (Spring Hill is only thirty-five miles south of downtown) and nearby Franklin (the seat of adjacent Williamson County, which is one of the wealthiest counties in the nation), plus its easy access to Interstate 65, this once-little town is still growing and sprawling, which drives more farmland loss.[6] The US Census Bureau lists the town's 2020 population at just over fifty thousand people, an astounding fifty-fold increase since 1980. Other

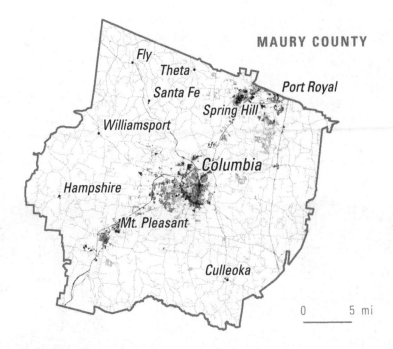

Select towns and communities within Maury County. (Map created by Luke D. Iverson; data sources: US Census Bureau TIGER, Natural Earth, National Land Cover Database 2016, Esri Living Atlas)

towns in Maury County have not grown quite so rapidly, but residential and commercial development—particularly near the county seat of Columbia and along major highways—has reduced agricultural acreage elsewhere too. Subdivisions, large-lot tracts, and strip malls now cover former fields, pastures, and forests.

Increasingly, the farmland that remains in Maury County—whether located in rural areas or adjacent to booming towns—is dominated by large-scale agriculture. From 2002 to 2017, the county's total number of farms declined from 1,754 to 1,583. During that same span, the average farm size in the county slightly increased by 7 acres,

A large new subdivision on former farmland between Spring Hill and Columbia. An ode—or, in the eyes of some, an insult—to the land that this development occupies, the name of this subdivision includes the word "Farms." (Photo by author)

from 137 to 144. As noted earlier, however, average acreage is not the best metric to analyze trends in changing farm size. A more detailed examination of data from the Census of Agriculture is needed.

In 2002, 997 farms between 50 and 499 acres were present in Maury County. These farms—which covered a total of 147,790 acres—helped support a smaller-scale, diversified, rural farm economy, though that localized economy was even stronger decades earlier. By 2017, the number of small and midsized farms declined to 761 farms that covered 113,958 acres. Drives throughout the county and observations

Large-lot residential tracts, or what some people call "mini-farms," built on land that was once a cattle farm in eastern Maury County. Along this rural, one-lane, chipseal road, over twenty homes have been built in the past few years, including some—like the white house in the middle of this photo—that are still under construction. Note the silo and barns in the background on the left, remnants of an agricultural history. (Photo by author)

from locals, which will be explored later, show that more small farms have been shuttered in the past few years.

During that same period, the number of farms between 1 and 49 acres—what I refer to as "tiny farms" or "mini-farms" in earlier chapters—actually increased, from 674 to 733 farms. Interestingly, the land area covered by these tiny farms dropped slightly, despite an up-tick of 59 farms.[7] In other words, these tiny farms—or, perhaps more

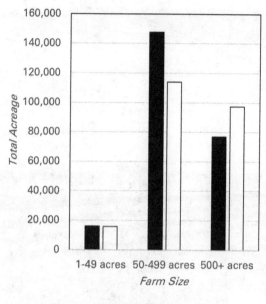

Total acreage covered by farm size in 2002 and 2017 in
Maury County. (Data source: US Census of Agriculture)

accurately, large-lot housing tracts or estates—became even tinier. Likewise, larger farms became even larger and more numerous. The number of farms over 500 acres in size grew by 6, from 83 to 89 farms. This increase is, of course, rather modest. But the explosion of acreage covered by large farms is astonishing. Whereas large farms occupied just under 76,900 acres in 2002, they came to cover 97,314 acres by 2017, an increase of 27 percent. And remember, this increase occurred over the same span when almost 14,000 acres of agricultural land were removed throughout the county. Reflecting larger trends across the state, region, and nation, the growing presence, power, and influence of large-scale farmers in Maury County reinforces the hardships that small and midsized farmers face.

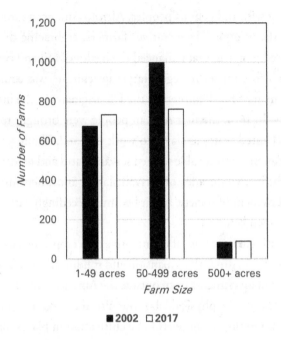

Total number of farms by size in 2002 and 2017 in Maury County. (Data source: US Census of Agriculture)

Reckoning with Racism in American Agriculture, Past and Present

The challenges confronting smaller-scale farmers in Maury County are significant. While the difficulties of dealing with farmland loss and agricultural consolidation are widespread and often intense, it's important to begin with a discussion of how race has impacted, and still impacts, Black farmers in this community and beyond.

For more than four centuries, racial injustice has been an insidious hallmark of US agriculture. People of color—particularly Black people—have long been treated unequally and oppressively, despite the invaluable knowledge and skills they have brought to the field. In fact, along with Indigenous peoples, African American farmers were some of the original "regenerative" farmers, embracing diversified, place-based practices that enhanced the health of soil and communities long before the term "regenerative agriculture" was coined. This mistreatment of Black farmers in the US originates in the institution of slavery. In 1619, the first African people were brought to what is now the United States as slaves to work the land. While a few early African Americans were able to shed servant status and tend their own farms within two decades of arrival, slavery morphed into an entrenched element of society, and it became exceedingly rare for Black people to own land.[8]

Increasing reliance on labor-intensive cash crops, such as tobacco, cotton, and rice, drove the expansion and solidification of US slavery, as did capital investments from outside the American South. Depending on not only the physical labor but also the expertise of enslaved people—for example, successful rice cultivation in places like South Carolina relied on the skills, experience, and knowledge of enslaved African Americans; evidence also shows that plantation owners depended on slaves to improve animal husbandry practices and to engineer successful infrastructure, like fences and canals—slave owners sought to further justify this wicked institution. Citing historical research by leading Black scholars, Waymon Hinson and Edward Robinson write that white aristocrats aggressively defended and promoted bondage. Among other false justifications, these planters claimed that Black people were inherently inferior to whites, argued that the entire South's economy necessitated slavery, and twisted certain passages from the Bible to make it seem as if slavery was permitted, even endorsed, by Christianity.[9]

Slavery continued and expanded through the eighteenth century, even as the nation's founders proclaimed inalienable rights to life, liberty, and the pursuit of happiness. By 1860, there were approximately four million enslaved people in the US South, all of whom were forced into various types of agricultural and nonagricultural labor. Although white people owned the vast majority of southern land, enslaved African Americans were largely the ones serving as stewards.[10]

With the Emancipation Proclamation, the Civil War—the outcome of which, historians have shown, was swayed by the courage of Black leaders and soldiers, women and men, who advocated and fought for their own freedom—and the passage of the Thirteenth, Fourteenth, and Fifteenth Amendments to the Constitution, slavery ended, and Black Americans gained increased autonomy. This advancement is certainly seen in agriculture. While many formerly enslaved people migrated to cities or continued to work as tenants or sharecroppers after the war, frequently on the same farms where they had toiled under bondage, others became landowners.[11]

One of the most famous examples of Black Americans securing land after the Civil War is represented by the well-known phrase "forty acres and a mule." In January 1865, technically before the Civil War had ended, Union General William Tecumseh Sherman issued Special Field Order Number 15, which stated that "not more than forty acres of tillable ground" be given "for the settlement of the negroes now made free by the acts of war and the proclamation of the President of the United States." The provision of a mule was not officially in the order, but the Union Army did loan or give some new Black landowners a draft animal to help them jump-start their farming operations. Although Sherman's order granted ownership of around 485,000 acres to Black Americans, it was geographically limited. Much of the land that was redistributed was near the Georgia and South Carolina coasts.[12] More expansive actions came later. In the Freedmen's Bureau Act of March 3, 1865, and the Southern Homestead Act of 1866, formerly enslaved people were given the chance to purchase land.

For a brief time, these initiatives helped African Americans secure their own farmland.[13] This progress, however, was soon erased. Because of intense political opposition, particularly from President Andrew Johnson, much of the land newly acquired by Black citizens was returned to wealthy white southerners.[14]

Stripping former slaves of their land was a grave injustice, one that denied many people the chance to start building generational wealth after lifetimes of bondage. As W. E. B. Du Bois wrote in 1935, "The slave went free; stood for a brief moment in the sun; then moved back again toward slavery."[15] Beyond having their land ripped away, violence toward Black Americans became rampant. White terrorist groups such as the Ku Klux Klan—founded in Giles County, Tennessee, just south of Maury County—as well as individuals and even government officials acting on their own, targeted Black people and white allies. In doing so, they sought to stifle liberty and retain hegemony. The Equal Justice Initiative speaks further about the horrors of this violent era:

> More than 4,400 African Americans were lynched across 20 states between the end of Reconstruction in 1877 and 1950. Racial terror lynchings were more than just hangings. They involved groups of white people committing acts of fatal violence against African Americans to instill fear in the entire Black community. These lynchings were frequently carried out in broad daylight, and perpetrators could expect impunity. Government officials frequently turned a blind eye or condoned the mob violence. This era of racial terrorism shaped the geographic, social, and economic conditions of African Americans, and America as a whole, in ways that are still evident today.[16]

As this quote shows, Black men, women, and children were murdered in states across the nation, with a large concentration of this violence in the South.

At the same time, racial segregation and Jim Crow laws became standard, and sharecropping increasingly became a new form of slavery. To be sure, sharecropping impacted farmers of all races, placing them in positions of poverty and little autonomy while demanding production. This exploitive arrangement is explored thoroughly via eyewitness accounts in James Agee and Walker Evans's classic book *Let Us Now Praise Famous Men*. But history shows that sharecropping was even more difficult for farmers of color. In Theodore Rosengarten's *All God's Dangers: The Life of Nate Shaw*, Shaw—a longtime Black farmer and sharecropper from Alabama—describes firsthand the lived experiences of many rural Black Americans during the late nineteenth and early twentieth centuries. The injustices that Shaw and so many others faced—and the extreme resistance and hatred they encountered when they stood up for themselves, their families, and their neighbors—are appalling. Fleeing these sorts of harsh conditions, many Black southerners moved to urban and rural areas in the North and West in search of better opportunities.[17]

Yet even in the face of lynchings, intimidation, and threats, a growing number of African Americans acquired and kept land in the South during Reconstruction and into the twentieth century (including the aforementioned Nate Shaw), occasionally with help from white neighbors or interracial relatives. Historical economic research indicates that, at least in some communities—including rural communities in Tennessee—Black farmers often had to pay more for comparable land than did white farmers, particularly when purchasing small farms. As Black farmers gradually began to cultivate their own land, leaders like Booker T. Washington, Ida B. Wells, George Washington Carver, and W. E. B. Du Bois—as well as lesser-known Black agrarian figures like Benjamin Hubert, Thomas Campbell, Robert Moton, J. W. Carter, and countless others—supported and advocated for them. With these leaders' help—and with individual courage, determination, effort, and expertise—Black farmers came to own over fifteen million acres of farmland by 1910. Jess Gilbert and her coauthors write that there were roughly 926,000 Black farmers in the United States by 1920, making up 14 percent of all American farmers. Then as now, most lived in the South.[18]

This early twentieth-century period represents a peak in Black farmers and farm ownership. Despite extraordinary efforts from Black Americans—efforts that included the enhancement of agricultural extension and education programs, establishment of agricultural cooperatives and rural support organizations, settlement of all-Black communities, development of interracial alliances, and more—Black landownership and agricultural involvements steadily declined. It is true that the number of farmers decreased across racial groups because of industrialization and economic trends. In particular, small farmers and farms suffered, especially after World War II when US industry and government turned wartime technology and techniques toward "modernizing" domestic agriculture, one of many factors that led the nation into today's exploitive, industrialized system. But African Americans experienced a much sharper decline than their white counterparts did. "Between 1920 and 1969 there was a 90 percent decrease [of Black farmers in the US], and by 1997 a 98 percent decrease," write Spencer Wood and Jess Gilbert. "This compares to an overall decline among white farmers of 66 percent."[19]

Some of this disproportionate decline can be attributed to Black farmers and their families proactively pursuing other opportunities, made possible by educational advancements or moves to urban areas. These changing social, vocational, and geographical patterns include what historians call the "Great Migration," an extended period from the 1910s to the 1970s when millions of African Americans left rural areas in the South for new homes throughout the nation. Decline can also be attributed to environmental factors, such as boll weevil infestations that destroyed cotton crops and pushed many smallholder farmers of all races off their land. Yet much of the decrease in farm ownership and agricultural involvement among African Americans since the early twentieth century is due to racism. As Gilbert and her colleagues state in a thorough literature review of the catalysts of Black farmland loss, "Racism stands out both by itself and as a contributing factor to many of the other causes of black land loss."[20]

In *Dispossession: Discrimination against African American Farmers in the Age of Civil Rights*, Pete Daniel explores these race-based causes of Black farmers' land loss. Many farmers, he explains, experienced prejudice and individualized racism, including verbal threats and physical violence. These actions took a massive toll. But it was more systemic factors that accelerated the widespread struggles of Black farmers. These structural injustices include racially driven delays and denials of loans to purchase land or maintain farm operations, reduced trainings and educational offerings for nonwhite farmers, and ignored complaints of racism made by Black farmers to USDA officials.[21]

Racial inequities also arose in extreme underrepresentation on local-level agricultural committees, which controlled and distributed many different forms of assistance for farmers. On this topic, Dorceta Taylor shares shocking statistics about the underrepresentation of Black farmers on county agricultural committees in the South. "Though the Southeast region has the largest concentration of black farmers," she writes, "in 1996 only 28, or 1.1 percent, of the 2,469 county commissioners in the region were black. . . . In fact, there were a total of 37 black county commissioners out of a total of 8,147 nationwide." This severe lack of representation harmed many farmers of color.[22]

In the final decades of the twentieth century, multiple government entities—including the US Commission on Civil Rights, the USDA's Office of Civil Rights Enforcement, and a specially formed

Civil Rights Action Team—documented persistent racism by the Farmers Home Administration, the Agricultural Stabilization and Conservation Service, the Farm Services Agency, and the entire USDA. Although some of these injustices were addressed by *Pigford v. Glickman* (1999, 2010), a two-part class action lawsuit that has provided over $2 billion for Black farm families that suffered racial discrimination at the hands of the US government, lasting damage had already been done.[23] As of 2017, Black Americans made up less than 1.5 percent of the nation's agricultural producers. They collectively farmed only 4.7 million acres of farmland, which equates to roughly 0.5 percent of all US agricultural acreage.[24]

In addition to the aforementioned persecution that thwarted Black farmers, the pervasive issue of heirs' property has also made it difficult for African Americans to retain possession of land. Heirs' property is "land held communally by family members of a landowner who has died intestate."[25] In other words, land becomes heirs' property when a current landowner dies without a clear will or estate plan, leaving land to family members in a somewhat unofficial and informal capacity. Frequently, these new co-owners are distantly related and may not live anywhere near the land in question, especially as property continues to pass from one generation to the next.

Sometimes called "tenancy in common," heirs' property affects people of all races, particularly families that have little wealth. It's common throughout distressed communities in Appalachia, for example, as well as other poor communities across the nation. But it is especially prevalent among African American landowners in the South. Experts state that the disproportionate impact on Black landowners can be attributed to several factors. For one, people of color have long lacked equal access to legal services, making it difficult to create valid wills. The lack of a clear and legitimate will can, as noted, quickly lead landownership into heirs' property status. Further, estate planning is a complex, time-consuming, and expensive process, one that can be hard to afford for many economically insecure farm families. And in some cases, people may choose to intentionally pass down land through heirs' property structures, believing that common family ownership protects the land and helps it stay in the family when, in reality, it can make folks vulnerable to exploitation. The factors that lead Black landowners and farmers into heirs' property status are rooted in historical inequities. But to be sure, these challenges are still present. Some estimates suggest that heirs' property characterizes ownership for well over a third of Black-owned land in the US South, and one study reveals that—because of heirs' property struggles, discriminatory lending practices, and other factors—Black Americans collectively lost $326 billion worth of agricultural acreage in the twentieth century alone, a massive loss of generational wealth that makes it harder for Black farmers to succeed in the present.[26]

While Janice Dyer and Conner Bailey argue that ownership through heirs' property can sometimes offer emotional and material benefits—and while owning land jointly with family members isn't inherently problematic—it can present severe problems under current legal structures. Without clear title to a parcel of land, it has been difficult, and at times even impossible, to use property as collateral for loans or to qualify for disaster relief. For farmers specifically, heirs' property ownership and the lack of clear title have historically prevented them from being eligible for conservation funding or farm financial assistance from the government. Efforts have been made to address some of these structural problems, which is important. Still, many Black farmers have already felt the pain of being excluded from these programs.[27]

The aforementioned heirs' property challenges are indeed very serious. But the most egregious injustice associated with heirs' property is related to dispossession and "forced partition sales." Describing partition sales, Gilbert and her coauthors show how heirs' property ownership frequently leads to farmland loss: "Any individual co-owner can legally demand his or her share of the property. This usually forces a sale of the landholding, regardless of the wishes of other heirs. Any heir may also sell a share to non-family members without consulting other heirs. Non-family members (usually white landowners) have used these laws to their advantage, buying a small share in a parcel of land in order to force a partition sale. They are then able to purchase the entire parcel, usually at a very low price." To build on this latter point about low prices, it is essential to note that forced partition sales often occur at public auction. This sale format offers limited notice to buyers and sellers and, as noted, frequently yields low sale prices, harming families. Sometimes, even just the threat of a forced partition sale—made by a deep-pocketed developer with experience in these predatory-but-legal practices—can be enough to force family members to sell their land.[28]

In the case of heirs' property, then, it only takes one person—whether a distant cousin, a developer who has bought into ownership, or another individual co-tenant—to decide the fate of a farm. Given the significant challenges that heirs' property poses for farmers, land tenure, and generational wealth, several entities and individuals have started specializing in serving people who hold land in this way. The Federation of Southern Cooperatives, the Black Family Land Trust, and the Center for Heirs' Property Preservation, among others, all work with people to remedy these title and ownership issues. The USDA has launched a program intended to help heirs' property farmers secure the funding needed to clear title to their land. And a recent legal effort called the Uniform Partition of Heirs' Property Act is aiming to address tenancy-in-common issues through state legislatures across the United States. All these efforts are helping move the needle toward progress and justice for heirs' property landowners. Yet even with these good and courageous efforts, heirs' property is still a major challenge for Black farmers. Lizzie Presser states that "the U.S. Department of Agriculture has recognized [heirs' property] as 'the leading cause of Black involuntary land loss,' " highlighting the pervasiveness of the problem.[29]

From heirs' property injustices and widespread violence to loan denials and rampant racism, Black farmers have faced tremendous difficulties throughout US history. Some of these challenges still exist today. And while it would be wrong to paint the story of Black agrarians as a purely declensionist narrative—historical examples such as Fannie Lou Hamer's Freedom Farm Cooperative and present-day endeavors like Leah Penniman's Soul Fire Farm and Michael Carter Jr.'s Africulture, among many other models, prove that farming can be a form of liberation for Black Americans—it is important to reckon with the reality of racism in agriculture.[30] Through the eyes of farmers from Maury County, several struggles are intimately revealed.

Samuel's understanding of the difficulties that Black farmers encounter started early. Although he spent his youngest years living in a different Middle Tennessee town, Samuel and his family moved to their farm in the late 1960s after his grandfather died. At that time, he was twelve years old. This land had been in his family for generations, and it was where Samuel's father was raised. It was also where Samuel learned his love of agriculture—and where he first conceived his now-realized dream of becoming a farmer. But telling this agricultural aspiration to his father elicited a somber and stern response. "I guess I was about fourteen or fifteen years old," Samuel said, "and my daddy, he was a mechanic. No offense now, but my daddy said, 'A Black man cannot make it in farming. What you need to be is a mechanic.' " Disappointed, Samuel protested. "I said, 'Daddy, I like farming!' " But Samuel's father, who anchored his advice in personal experience as a part-time farmer, was firm. "He said, 'Son, I'm telling you— a Black man can't make it.' " Heeding his father yet still holding out hope to one day own and care for a farm, Samuel took auto-mechanics classes in high school, but he also participated in Future Farmers of America.

While Samuel's father experienced direct, personal racism in agriculture—for example, Samuel shared that his dad was sometimes grossly underpaid compared to fellow white workers when he worked for other farmers as a young man—structural discrimination proved more damaging. To supplement the preceding historical literature with personal experience, Samuel explained that securing loans was particularly problematic. "Back then, a Black man couldn't get a loan [to support his farm]," he explained. "It was hard for him to get a loan to get something. My daddy tried it. When we moved back [to the farm] . . ., he was going to try to get our homeplace set up pretty good to work. He couldn't get no kind of loan." Though Samuel's father was eventually able to purchase a tractor and some implements for the farm by working his way up as a mechanic at a nearby factory, farm-support loans would have made success quicker and easier. For some families, this inability to access funds led to dispossession.

Others shared similar stories about securing loans. Ronald and Walter are brothers, and in each of our interviews, they both highlighted borrowing money as an issue affecting Black farmers. Ronald, who does most of the farming in his family, explained that landing agricultural loans—while not as difficult as it once was—is still an issue for Black farmers. Noting that he has the same credit score as many of his white farming friends, Ronald expressed doubt that he would be approved on the spot for a loan to buy a new tractor at a dealer. He feels confident that his friends, on the other hand, could "go up there and buy a $70,000 tractor all day." Walter, who helps out on the farm with manual labor as needed and visits the land often, affirmed his brother's comments. "The number-one thing that's affecting Black farmers is getting access to funds. And I mean as a loan—to get farm equipment and to better farm," he stated. "Most of the [loan opportunities] I started finding out about over the last fifteen years that would benefit farmers, I found out about them at forty [years old]. My uncle's seventy, and he was just finding out about them. . . . That knowledge was still not getting to the people it needed to get to."

Further emphasizing loan problems, Ruth—a well-known community leader who now lives in Columbia but still co-owns and tends her family's land—posed a set of exasperated rhetorical questions in response to a query about ways to support Black farmers. Describing the various difficulties confronting most people who want to get a start in agriculture, she asked, "For somebody today [to buy a farm]? . . . And a Black farmer too? Who's going to loan you that money?"

Participants' experiences with systemic injustice are not confined to farm loans. Limited access to educational resources was also cited as an impediment to the success and survival of Black farmers. Walter stated that Black farmers have often not been shown certain techniques and advancements that increase yields and profits, especially in past decades. These farmers had deep localized knowledge of their land, but new and helpful trends were often slow to be shared, making it harder for struggling farmers to stay afloat.

Some historically Black colleges and universities, like nearby Tennessee State University (TSU) in Nashville, offer educational materials and extension services specifically for Black farmers. These institutions also focus many extension efforts on serving small farmers, which is critical since it's statistically more likely for farmers of color and other underserved populations to be operating on small and midsized scales. Sometimes called "1890s schools" because of their creation via the second Morrill Act in that year, requiring segregationist states to either integrate their land-grant colleges or provide separate schools for people of color, schools like Tennessee State make these efforts despite being chronically underfunded. For example, a recent investigation uncovered that the state of Tennessee withheld funds from TSU for decades, shorting the institution by between $150 million and $545 million. Allotted-but-elusive funding could have helped the university "offer more competitive scholarships, pursue more innovative research, and invest in extension programs," according to TSU's president. This sort of underfunding—or what some experts have simply called "theft"—is not unique to Tennessee. An investigation by *Forbes* analyzing per-student funding levels suggests that between 1987 and 2020, the eighteen historically Black 1890s schools were underfunded by an aggregate of $12.8 billion, adjusted for inflation.[31]

The federal government (including the USDA), state agricultural departments, and flagship university extension programs have stepped up efforts to support Black farmers in recent years.[32] Yet Ronald feels that he and his peers are still left out. For example, he shared that he knows white farmers in the community who have been taught how to navigate farm finances and leverage assistance programs, making it easier and cheaper to secure much-needed resources and equipment. With a frustrated laugh, he said, "They don't teach Black farmers how to do that! They keep that under their hat."

Part of the educational and service disconnect is due to justifiable mistrust of government programs from Black farmers themselves. The government has long marginalized nonwhite farmers. This historical mistreatment has led many Black farmers, especially those who are older, to steer clear of farm-support programs, even those specifically designed to help them. "Older Black people have always been taught, 'Don't let the government know what your business is,' " Ronald explained. "And they're scared to go talk to these people [government and extension officials]. . . . I guess they're just hesitant on how much information and business they're willing to share with the government." Ronald believes that farm-service providers and agricultural officials must make more of an effort to go out into Black communities and do proactive outreach, rather than waiting for these farmers to walk into an office. Doing so might address some of the gaps in service that exist, especially if more people of color can inhabit leadership roles in the farm-service sector.

The final structural challenge related to race that interviewees mentioned is landownership through heirs' property. As in other communities throughout the South and the nation, this type of ownership affects Black farmers in Maury County. Ronald told an illustrative story. A will must be prepared—and prepared "exactly right," he said—or else "somebody will come and weasel their way in and find a loophole to either sell that land or force it to be sold." In a personal example, he showed how heirs' property can lead to dispossession and despair:

> We've got a 200-acre farm out there. But the majority of it belonged to my uncle, and he passed away a couple years ago. Well, he owned about 120-something acres out of that farm, and then the rest of it was heirs' property. But he was the one who had maintained all the taxes and all that good stuff, and nobody really knows where the property lines are. We have an idea of where they are, but . . . It's one of those deals where if something don't happen in the next few years—with getting it into some sort of trust where it can't be sold outside the family—eventually it'll be sold. In our particular situation, he [my uncle] didn't have any kids. OK? Instead of him having the knowledge to have the will drawn up like it needed to be, he drew the will up leaving it [the land] to all to his brothers and sisters. The problem with that is all his brothers and sisters are old! So then what happens is [when they die], their part of the property goes to other family members and stuff. . . . I've got cousins in South Carolina and different places that ain't never set foot on that property.

After hearing this story, I ventured a follow-up, clarifying question: "Now your cousins are about to be part owners of the land?" Yes, Ronald answered. "And what do you think is the first thing they're going to want to do?" he asked in return. Remembering that the cousins had never even seen Ronald and Walter's family land, I offered a cautious guess: "I imagine they might want to sell it?" Ronald boomed, "That's exactly right!" Noting the million-plus dollar value of the land, he fears that the farm is in trouble.

These various sorts of financial and landownership trauma lead to great stress for Black farmers, a feeling most interviewees hinted at or mentioned explicitly during our conversations. That stress alone can be devastating. But unsurprisingly, chronic stress, anxiety, and depression lead to other health and well-being problems. Needing to maintain a constant defensive stance and always worrying about what could happen to cherished family land can lead to physical ailments like high blood pressure, increased heart rate, insomnia, heart attacks, nervous breakdowns, and shortened life expectancy. Some health experts contend that these and other mental and physical health factors constitute a public-health crisis among Black farmers specifically.[33]

In addition to the aforementioned systemic challenges and the stress that they cause, people mentioned individual experiences with discrimination too. Unlike historical examples cited earlier, no individuals described experiences with verbal or physical violence. Instead, they noted awkward situations and offhand remarks made by coworkers, neighbors, friends, and strangers that question their involvement in agriculture. These subtle comments—often known as "microaggressions"—add up over time and cause anger, frustration, and pain.[34]

Ruth shared two illuminating stories. In the first, she mentioned attending the annual Farm Bureau breakfast in Columbia, where local farmers are honored. She often receives curious looks while there, as if she does not belong. Few people seem to believe that her family has been farming the same land for over one hundred years. In the second story, she described a recent experience with a colleague. Ruth serves on the board of a local nonprofit, and this organization held an auction to raise funds for its work. During the event, which was held virtually because of COVID-19 public-health concerns, Ruth made a comment about the nature of the bidding and the sale. Reflecting on fond childhood farming memories, Ruth typed into the chat, "Oh my, this reminds me of the cattle auction we used to go to!" Immediately, she received a response from a fellow board member who could not believe her agriculture-related comments. "Ruth," the man said, "you don't know nothing about farming!" Others laughed. Ruth is tough and resilient, but these experiences upset her. Through the phone, I could hear the hurt in her voice. She is proud of her family's agricultural heritage. To have it questioned is wearisome.

Ruth's youngest brother, George, referenced similar experiences. After years of working with some of the same people at his full-time job in Nashville, he still gets peculiar looks when he talks about farming on his family's land. "Any [farming] story that comes out of my mouth, they kind of look at me like, 'Yeah, I don't know if that's true or not,' " he said. "They can't picture me in overalls and rubber boots because I'm sloshing in the mud to put hay out for cows."

Going beyond this theme of microaggressions, Walter expressed frustration about disrespect for Black farmers in general. "A lot of folks— a lot of African American people even—don't respect Black farmers," he said. When he speaks with others about race, rural communities, and agriculture, he discusses disrespect through misrepresentation:

> One of the things I've always brought up is in movies and television and mass media, you always see an urban Black person. You always see a Black person that's maybe in some type of jet-set lifestyle or some type of athlete. But you never see an accurate portrayal of a Black farmer on television. And that's where everybody came from! . . . If you do see that, it's in some oppressive type of slavery narrative instead of the situation my grandparents grew up in and my mother grew up in. It was agriculturally based, but it wasn't oppressive. It was

something we did. We made money off it, and, you know, it was a pretty good living—just a little harder than everybody else. But they survived doing it.

Depictions that are perpetually and purely oppressive, Walter believes, discourage young Black people from pursuing agriculture as a vocation. "We have to make a positive portrayal of agriculture to African Americans. And that's going to have to come from us. That's the reason why I try to be an ambassador for agriculture in my school and try to get more African American kids to take the [agricultural] classes." With an agrarian ethos, Walter argued that "everybody needs to know how to grow something."

Despite these discussions of structural racism, microaggressions, and misrepresentation, multiple farmers also explicitly described healthy race relationships in their rural communities. Pointing mostly at systemic problems as the culprit for Black farmers' struggles, Walter said that, in his personal experience, "the individual relationships between white and Black people are very good." Likewise, Samuel mentioned having white neighbors who were eager to help him in a pinch. When some nearby farmers learned that Samuel had tractor trouble a few years ago, they jumped in to lend a hand. "They'd come to me— they'd hear I was down—and say, 'Use my tractor and then fix yours!' So I've got good neighbors. . . . I probably got the best neighbors around." These stories offer evidence of and hope for kinder and more equitable rural communities, especially if the same kind of individual respect, empathy, and solidarity can be translated into more just structures and systems.

Still, the factors discussed in this chapter have had a heavy impact on Black farmers in Maury County and beyond. Whether because people voluntarily pursued nonagricultural opportunities or were forced out of a life and livelihood they loved, a fact remains. As Ruth noted during our interview, "There's just not that many farming Black people in Maury County."

Gutted Ground: Watching Agricultural Acreage Disappear

In addition to the race-based challenges just described, interviewees also cited the loss of farmland from development as a difficulty for small and midsized farmers in Maury County. Over the years, they have watched much of the county transition from formerly rural communities into suburbia. In a concise but telling statement, Walter explained that, throughout a large chunk of the county, "you went from farmland to urban sprawl."

Aligning with statistics from the Census of Agriculture, multiple interviewees pinned amplified development and farmland loss in Maury County to the opening of the General Motors automotive plant. When I asked Ruth if she had noticed many changes in the county's landscape in her lifetime, she answered, "That would be a yes. Thirty years ago, when Saturn came into Spring Hill, Tennessee, it changed everything." George and Walter echoed Ruth's comments, as did Ronald. "One of the big issues that has happened to farmland—the loss in Maury County—is from when the General Motors plant came to this area back in the late '80s," Ronald said. "People transferred here from different areas, and they were offering top dollar to farmers. Some big builders were coming in and putting in these big subdivisions to increase the population, to build all these houses for the people who came in here with that plant."

People certainly swarmed Spring Hill and surrounding communities in the 1990s. Population data cited earlier is proof of that. And they keep coming, especially given the town's proximity to Nashville. On this continued population growth, Walter shared an insider's perspective. In 2005, he was working on a graduate degree in the hopes of becoming a school principal. The director of the county school system invited Walter to attend a special planning meeting, saying that—given his future vocational goals—he needed to hear the information that would be shared. During the meeting, an official from a federal agency pulled up a map of Maury County and then turned to address the small crowd of local leaders. "Gentlemen and ladies, I'm calling you in here tonight because I'm about to retire, and I've got a dire warning for Maury County," Walter remembered the man saying. The warning was aimed at population growth and infrastructure needs. Over the next twenty years, tens of thousands of people would move into the areas around Spring Hill and Columbia, the official explained. Current infrastructure—schools, roads, sewers, emergency services, treatment plants, and more—could not handle that intense growth. The county would incur massive costs, and the community would be forever changed. "Everything that this fellow said became true," Walter lamented.

Offering another angle to this story, Walter shared that he now works a second job to bring in extra income for his family, and in this role, he often interacts with out-of-town visitors. Through conversations with folks while on the job, he has learned that many people are moving into Maury County from out of state. Some growth is fueled by locals, but a lot of people are "moving here to escape Illinois, Michigan, Minnesota, Ohio, Indiana, Wisconsin": "All those states I've mentioned, they're moving here to escape." People are coming from places like California too, as a Robertson County farmer noted in chapter 3. A *Los Angeles Times* article notes that "nearly 12,000 people left California for Tennessee in 2019, according to US Census Bureau data, up from roughly 9,600 in 2018 and about 7,900 in 2017." Most moved to the Middle Tennessee area, including Maury County.[35]

Many of these new residents, Walter has observed, are older. Noting Tennessee's lack of a state income tax and its mostly favorable weather conditions, he said, "They feel like if they move here in retirement, they can maximize their retirement and have more money in retirement than they would have if they were still in the North. And it's warm down here. So that's what we're dealing with right now. . . . And it's not a good thing, in my opinion, because it's eating up farmland." Retirees may also be moving to the area to live closer to family members who came to Maury County and nearby Nashville for new jobs, such as in the automotive, tech, and health-care industries.[36]

Building on these observations, Walter shared that residential development—rather than commercial or industrial—is the primary driver of farmland loss in the community. When I asked him what is taking the place of farmland, he quickly said, "Houses, housing developments, subdivisions, and five-acre lots." Walter soon augmented this answer. "Now, you're seeing the fast-up development and the ticky-tacky houses—in other words, the same-looking house in the same place.[37] And the greatest example of this again is if you go into Spring Hill." To illustrate the point, Walter spoke of a specific farm that he used to love driving past on his way to Spring Hill. It was "one of the legendary farms in Maury County that even up until fifteen or twenty years ago, it was producing honey and a lot of pork, especially ham. And that's gone now. Huge subdivisions are being developed on it as we speak. It's gone."

Like Walter, Samuel offered a unique perspective on the development occurring in the county. Noting his full-time job as a mechanic, he said, "I work on heavy equipment and stuff, and I go where things break down." While he has not worked on many dozers or dump trucks in the rural area immediately around his farm, he has made lots of equipment repairs on former farmland in other parts of the county. "About eight to ten miles, I'd say, east of us, they started cutting these farms up into subdivisions. I hate that it happened. They're taking good land and making subdivisions out of it when they need to take these fields and stuff that you can't cultivate and make subdivisions

A farm for sale on the outskirts of Columbia. Note the wooden barn—which may soon be demolished—in the far left background. (Photo by author)

out of them." As in Robertson County, it seems that the best farm-land is sacrificed first. "Up around close to the city limits," Samuel continued, "they'll take a hundred-acre farm and put seventy, eighty houses on it. And it's just gutting the farmland."

Taking a drive through the county confirms that subdivisions are spreading, but larger-lot residential development is rampant in Maury County too. While Samuel noted that farmland is not being fre-quently carved up in his hilly corner of Maury County, others shared that real estate development has made its way into their once-rural locales. Ruth remembered aloud when her family's farm was still on a dead-end chert road in eastern Maury County, but nowadays, she noted, "There's a lot of development out there around us. They're building $400,000 to $500,000 houses around our farm." These new homes can inflate adjacent property values, making it more expensive for longtime rural residents to stay in place. And when developers build these new homes, they are too often unaware of—or apathetic to—the concerns of nearby property owners. When new homes started shooting up around her, Ruth recalled,

> They [developers] fussed at me and came into my job like five or six times without notification and without making an appointment through my secretary or anything. They would just show up at my office and fuss at me about getting access to the electrical pole that was on my property. That was the only way they could build [these houses]. And they never offered me a dime for coming on my property and moving that pole and placing another one. . . . They were just worried about building those nineteen houses—it's twenty-four total, but they built something close to nineteen of them then. They were just worried about that. It kind of fretted me. I wasn't going to hold [them] up, but it's like . . . You're paying everybody else for stuff, but the fact that we own this land is of no value to them.

George, building on his sister's comments, said that some of these new nearby developments are striving to be miniature farms. To the detriment of the land and the livestock, people will take a few acres in front of their homes, put up fencing, and bring in several horses or goats, he said. Grasses are overgrazed, animals are underfed, and soils are left bare. Heavy rains then take topsoil and send it washing into ditches and streams. George watches it all happen from atop his tractor across the road.

Further, the close-knit feel that used to characterize the area is gone. With a touch of nostalgia, George said that, years ago, people used to wave to one another when they passed on the roads. His new

The telltale signs of farmland conversion are visible in this photo. Stakes and posts indicate where land will be divided. New culverts and crossings will soon ferry in construction machinery. Although difficult to distinguish in this black-and-white image, the sun is setting on this once-healthy hay field.
(Photo by author)

neighbors do not appear to have gotten the message. "Now, nobody waves," he said. According to George, some people look at him in wonder and confusion as they drive past, as if they are surprised to see someone, especially a Black man, on a tractor mowing a hillside field. Other interviewees affirmed some of these siblings' comments about large-lot developments and cultural change.

Trying to make sense of what they're seeing, farmers offered various reasons for the substantial farmland loss occurring in Maury

County. For the most part, these reasons mirrored those that people in Robertson County shared. For one, land prices have increased significantly in much of Maury County, leading some people—especially if they are struggling financially—to sell their farms. As Ronald noted earlier, a two-hundred-acre farm could easily bring millions of dollars in instant income, much-needed money for many working-class rural families. A quick online search for "land for sale in Maury County, Tennessee" further illustrates the high value of undeveloped land. Every participant mentioned that these high values especially lure younger generations into selling recently inherited farmland. When children leave the farm and start pursuing nonagricultural careers, they often lose their connection with the farm, Ronald said. After parents or grandparents who have long cared for the land pass away, he continued, "there's nobody there to continue doing the farming, and then the farm ends up selling eventually."

In other cases, several farmers stated that a diminished work ethic was responsible for many young people selling family land. Farming is hard work, and not everyone is up to the task. Many next-generation farmers have "probably seen their parents struggle to farm," and they do not want to take on that same struggle, Ronald said. In another representative comment, George elaborated, "When you look at ninety-two degrees and being outside in the elements versus an assembly line or any other type of job you could have that's gonna allow you to be in from the elements during a portion of your shift, if you will, . . . the comparison in some people's minds is not even a comparison." Continuing his discussion of farm labor, George said, "If a cow is calving and that calf is not coming out and there are some issues going on, you've got to stay with that cow until that calf comes out. And that might be nine thirty at night or whatever. You can't say, 'I'll do it tomorrow' or 'My shift is over. I'm going home.' It doesn't work that way." Emphasizing the constant dedication it takes to farm well, George noted that there is little downtime for a farmer, especially one who also works a full-time off-farm job. Far from being a scrooge or a cynic, George and others just felt that many young people, including some in their own families, do not want that constant responsibility.

But as in Robertson County, blaming the loss of farmland on the idleness or indifference of young people is not entirely fair or accurate. While it's true that many may lack the desire to continue a family farming legacy, there are plenty of people in the next generation who dream of caring for a farm, whether they grew up in agrarian settings or not. For lots of folks, that dream is simply out of reach. Forced to compete with developers who have deeper pockets—and knowing that small-scale farming does not usually promise sizable economic returns, at least given our current agroeconomic climate—young, wannabe farmers are often outgunned on the land market. The widespread development of farmland, Ronald explained, "has thrown the prices of land up so high around here that a lot of people couldn't even afford to buy a farm if they wanted to." Building on Ronald's observation, Ruth said, "If you're talking about somebody having to buy 110 acres [the approximate size of her family's farm] in Maury County? I mean, there's just no way you can do that. You start out in the hole."

Data from other parts of the nation affirm their observations. In a 2022 article titled "Beginning Farmers, Farmers of Color Outbid as Farmland Prices Soar," Greta Moran writes that farmland has consistently increased in value—and increased sharply in recent years, especially with people fleeing urban areas during the COVID-19 pandemic—making it hard for farming-focused folks to buy land. Yet it's not just developers, industrial farmers, and local buyers who are driving up prices and outbidding aspiring small and midsized farmers. Across the nation, farmland has become an investment asset. As of 2022, Bill Gates is the largest individual owner of farmland in the US. Institutional investors like TIAA and Prudential have gotten into farmland ownership too, seeing it as a way to earn strong and consistent returns. Madeleine Fairbairn discusses the financialization of farmland in her award-wining book *Fields of Gold*, showing that this trend has major implications for rural communities and economic justice. Even if land owned by investment companies, Gates, and other billionaires is leased to farmers and remains in agricultural production, it still presents a major problem. Land—the foundation of farming—is being consolidated into fewer and fewer hands.[38]

Whatever the reasons that spur farmland loss in Maury County, the erasure of agricultural space causes emotional distress for farmers. All participants shared that watching development overtake farms in their community is difficult. On this phenomenon, George stated, "It hurts me from the inside." Likewise, his sister Ruth said, "It's just heartbreaking that farmers can't make it now. . . . It's just horrible that we're losing that land." Using similar language, Samuel said that farmland loss "really hurts" him, especially given his longtime love of agriculture. As with farmers in Robertson County, the clustering emotions of hate, hurt, and heartbreak were common reactions to farmland loss.

Other emotions were mentioned too. Ronald explained that seeing farms transform into residential developments "really upsets" him: "I don't think people realize that if it wasn't for farmers, eventually we won't have nothing to eat. But yeah, it's disturbing because it seems like everybody just wants to sell off and build, build, build. They're not thinking of the big picture." Walter echoed his brother's reactions, saying, "Oh, it's upsetting to me." Commenting on the explosive growth of a community in eastern Maury County that he has long been familiar with, he elaborated in a gloomy tone:

New homes in the once-rural Port Royal area close to the county line. For some people, like Walter, who knew this land well, seeing its transformation is especially difficult. (Photo by author)

It's upsetting because you knew the families that owned those farms, and for whatever reason—regardless of whether or not it was debt or whatever happened—those people don't own that land anymore, and it's not being tended to as a farm. You know, you used to be able to go and see cattle on that land, and now it's houses. It's a culture shock. . . . You were talking about emotion. I went through the Port Royal area of Spring Hill, that area. We used to hunt out there! Yeah, we used to hunt out there. And I was going through, and you can still make out the lay of the land. . . . And it's like, "Wow, we used to hunt out here." Now I can't hunt no more, but there's houses. And you get all of these [new houses], and it looks like I'm in New York City or Philadelphia or something. But it's the "country." It's confusing.

Paving the way for more development on what was once farmland near Spring Hill. (Photo by author)

Repeated trips to the Port Royal area might lead others to share in Walter's shock and confusion. On multiple recent occasions, I have driven through the same once-rural area he described. On one trip, I passed a field dotted with round bales of hay. Six months later, that same field was covered with new construction. Frequent news reports that announce the coming of hundreds of new homes show that this trend will continue.

The emotional trauma brought on by farmland loss weighs on people who are still trying to be stewards of the land. No matter the reason for reduced agricultural acreage, the dwindling number of farms and farmland tempts participants with despair.

The Big Getting Bigger: Farm Consolidation

In the same way that interviewees have witnessed the prevalence of farmland loss, they have also observed the growing influence and footprint of large farms in Maury County. Although their comments about agricultural consolidation were, on the whole, more subdued than their remarks about farmland loss, all participants noted the impacts that large farmers and flawed systems are having on smaller-scale farmers.

These impacts are illustrated through personal stories. As in Robertson County, tobacco production offers a helpful example of the struggles faced by small farmers. All five interviewees mentioned raising tobacco on their farms in the past. Decades ago, growing a few acres of tobacco—often burley tobacco in southern Middle Tennessee, as opposed to the dark-fired variety grown near the Kentucky border—could keep a farm financially afloat. "Years ago, you could make a profit and a living with tobacco," Ronald said. It was hard work, but it paid the bills and gave families some spending money. Ruth remembered thinking of her Christmas presents while working in tobacco as a child. Knowing that the quality of their family's crop influenced how nice their toys would be, Ruth and her siblings summoned extra motivation to do good work. Similarly, Samuel took pride in his family's tobacco-raising abilities. When it came to cutting tobacco and hanging it in barns—which is hot, sweaty, grueling work—few people could keep up with him and his brothers. This good work led to self-satisfaction and a strong reputation, as well as farm- and family-sustaining monetary value.

Today, the Census of Agriculture lists only one farm that grew tobacco in Maury County as of 2017. Mentioning government policy changes, such as the buyout program discussed in chapter 3 and the closure of the tobacco warehouse that used to be located in Columbia, Ronald explained that an important revenue source for small farmers disappeared when raising tobacco was no longer a viable option. While decreased tobacco use is undeniably good for the health of Americans—tobacco products are, after all, widely known as addictive, harmful, and carcinogenic—the fact that remaining tobacco production is dominated by large farmers is tough on smallholders in Middle Tennessee and elsewhere.[39]

Beef-cattle production—which is the most popular farming practice in Maury County, according to the Census of Agriculture—also offers an illustrative case. Walter, George, and Ruth all spoke briefly about the difficulties facing small-scale cattle farmers, but Ronald and Samuel offered the most revealing insights. Speaking specifically about the economics of raising beef cattle, Ronald said, "It seems like the costs of everything—equipment, diesel fuel, everything—keeps going up. And then the profit margin isn't as big as it needs to be. It's starting to be one of them deals where . . . it's almost like if you're not raising two hundred head of cattle or three hundred head of cattle"— in other words, if you are not a fairly big farmer—"it's almost come to the point where it isn't really worth fooling with."

Samuel also commented on the financial struggles he faces as a livestock producer. On his farm, he raises several different kinds of animals, including cattle, goats, horses, and mules.[40] In recent years, feeding his family's diverse livestock through the winter has become more difficult and more expensive. That's largely because of the expansion of large farms, he said. Whereas Samuel used to rent some nearby land to graze livestock and grow hay, he can no longer compete with huge row-crop farmers who—because other crop lands have been lost to development and because big farms keep expanding— have started leasing these places themselves. They can pay a higher rental price than farmers like Samuel can, so they gain access to the ground. "A lot of these here grazing lands, people are renting it for soybeans, which a cattle farmer can't pay $200 an acre for grazing land. The grain farmers are paying $200 an acre for it, so that's cutting back on your small farmers that used to rent this land to raise their cattle, sheep, goats, and stuff." Expanding on this discussion, Samuel went on to say,

> A small farmer's having a hard time making it right now because they can't get the land to raise hay, corn, and stuff [for feed] because all these big farmers are coming in, renting all these farms that little farmers used to rent. Say, five years ago, a fifty- or sixty-acre farm, one of these big grain farmers wouldn't even look at it. And the little farmers would rent it, raising hay or corn or something to feed their stock. But now, these big farmers are renting all the land. If they can get their combine on it, they'll rent it. And that cuts the little farmer out.

Because "you can't afford to rent that land to bale hay"—and because what land Samuel could afford to rent was twenty miles away, making it difficult to get the needed equipment to the field—he and his wife were forced to start buying hay from other farmers. This means that Samuel's hay equipment sits idle during the summers, which disappoints him in both personal and financial senses. He likes cutting, raking, and baling hay but just can't afford to do it. When competitors can easily outbid him and pull onto rented land with millions of dollars in equipment, Samuel has little choice in the matter.

Across the county and the country, other small and midsized farmers face difficulties like those described in this chapter. In a short but powerful statement, Walter summarized the situation for his fellow small farmers. "The little farmer—white, Black, indifferent—they're struggling."

6

"I Wouldn't Take Nothing for It": Stewardship Virtues and Place-Based Perseverance

Quantitative data, historical context, and personal experiences highlight the difficulties that smaller-scale Black farmers in Maury County face. As these multiple sources show, remaining on the land has been anything but easy. It would have been simple, understandable, perhaps even logical for them to abandon agrarian lifestyles and sell their farms. They could have cashed out and never looked back. Instead, these farmers have persisted in place, demonstrating deep and enduring commitments to caring for the land. The stewardship virtues they have nurtured over years, decades, and generations sustain these commitments.

Black farmers in Maury County practice imagination, affection, and fidelity in very similar ways to white farmers in Robertson County, revealing mutual loyalty to the land. Other supplementary stewardship motivations, such as religion, family, and economics, were also discussed in parallel manners by the people I interviewed. Yet one crucial difference exists. Black farmers in Maury County spoke explicitly about forging their own sense of belonging in agricultural communities, despite how the wider world tends to see—or ignore—rural Black people. These folks are acutely aware of the small population of Black farmers in their county, state, region, and nation. Defying the odds, weathering injustices, and continuing to farm is a point of deep pride for them. For them, presence is particularly powerful.

Imagination

Imagination—or an intimate and evolving understanding of, familiarity with, and attunement to a specific place—was a consistent theme across all interviews and the entire grounded research experience. In various ways, every farmer explained how they have cultivated and practiced this foundational virtue. They all started developing imagination as children.

George and Ruth attribute the beginnings of their hard-earned imagination to childhood chores. "Everybody had a role," George explained, speaking of the work he and all his siblings used to do while growing up on the farm. As the youngest, his tasks started off as simple. For example, he made countless trips to the chicken house to gather eggs as a boy. Though his chores might have been a bit easier to execute than the work taken on by older siblings, George was proud that he was contributing to the farm even in his earliest years.

As he grew and his older siblings moved out of the family home, George's responsibilities increased. He recalled feeding square bales of hay to cattle twice a day during the winter, regardless of the weather conditions. "It didn't matter if it was raining or snowing or cold or whatever. Those cows knew to come up in the morning because they were going to get fed. Then," he laughed, "in the evening, 'Oh, guess what—George is going to put out some more hay when he gets home from school!' And they [the cattle] would already be there." He took this work seriously, saying that—even on days he was physically exhausted—"it was something you couldn't avoid."

Ruth took her work seriously too and affirmed that every person on the farm, siblings and parents, had an essential role to fill—though some siblings enjoyed the work more than others. They usually woke up early to do their morning chores. Once the work was done, the family ate breakfast together, and then they left the farm. Ruth and her siblings walked down the road to catch the bus for school, and their parents—their father worked at a factory, and their mother was a nurse—left for work. Many days, she explained, "We'd get back in the afternoon, walk that mile [from the bus stop to the house], put our books down, and go to the field. We worked in the field until we got a break at five thirty or six [p.m.], and then we'd come in and eat supper, get our lessons, take our baths, and then go to bed and start all over again."

These active habits of tending the farm, both in the past and in the present, helped Ruth and George develop intimacy with their family's place. Both described still feeling deep attachments to the land today. Like farmers in Robertson County, they see the land as an extension of themselves. George, who lives in town but frequently drives to the farm to mow fields and check fences, said, "I feel really funny if I'm not there every couple of days. I mean, it's just a part of me." Reflecting on a recent experience, he continued, "Just the other day when I was there, a storm came up. And I just sat out there in the rain under a tree in a chair—because that's something that I did forty-five years ago. . . . I sat there with the rain droplets falling through the tree on me and just absorbed it as it was happening." In the same way that the rain nourished the family's beloved ground, it also gave life to George.

Recalling other farm memories—selling pigs, growing vegetables, hauling hay—and describing how they inspire him still, George said, "All that stuff, from the very beginning, is just inside me. And I wouldn't trade it for nothing." In a short-but-sweet affirmation, Ruth jumped in, saying, "We feel the same way about it!"

Like George and Ruth, Ronald and Walter also trace their imaginations back to childhood. When I asked Ronald how long he has been involved in farming, he said it is a lifelong endeavor. "Pretty much ever since I was born, I was involved in some kind of farming," he answered. "But when I was eight years old, I got introduced to the dairy side of farming." At that time, he started working with some relatives on a different farm. After school and on the weekends, Ronald would milk and feed cows, clean barns, and more. His relatives offered him a gift to show their appreciation for his good work. "They gave me my first cows—they were Holsteins," he said. Ronald kept those cows and, over time, has crossbred them and their offspring with Angus bulls. Decades later, Ronald's current beef herd owes its origins to the handful of Holsteins he earned as a child. The cattle, like Ronald, have a history on the land, one anchored in familiarity.

While Ronald mentioned taking classes in recent years to learn more about testing soils and fertilizing land, he attributes most of his farming knowledge to place-based experiences like those just described. Through spending time on the place, he has learned "what's going to grow and where it's going to grow and where your best stuff is. We know where our best fields are, and we know pretty much what needs to be planted or fertilized or what needs to be put on the ground." Also speaking about the power of good exemplars, he described where his agricultural intelligence comes from: "Being around farmers for most of my life and learning from my uncles and stuff growing up, watching them and following suit with how they had done things over the years—that's pretty much where that knowledge comes from."

Walter also noted that learning from others was essential for sharpening his imagination of the farm. Passed-down oral histories from older family members—grandparents and parents, aunts and uncles telling stories while stripping tobacco, working cattle, and sitting on front porches—have been particularly important for deepening his understanding and attachment. It was through these stories that he first realized the long connection his family has had to the land. "At least five generations have come through here," he shared with pride. "That farmland has been in our family since the patriarch of our family . . . bought his way out of slavery in the early 1860s." Thinking of the many family members who have cared for this place, Walter said that spending time on the farm "grounds" him. Through understanding the past, paying attention in the present, and envision- ing a hopeful future, Walter's imagination becomes even richer and more focused.

Aligning with others' comments, Samuel said that his farming knowledge is anchored in childhood experiences. "I've been around farming all my life," he stated early in our conversation. Expanding on this point later on, he explained,

> When we were big enough to pick peas and to chop tobacco and stuff, our daddy had us in the field. Fixing fences, our daddy had us out there. We were probably in the way, but he had us out there, teaching us to do this and that. . . . And the way we got the knowl- edge of farm work, we done it all hands-on. And that's where I learned most of it. I learned a lot at FFA [Future Farmers of Amer- ica], but I knew a lot before I got there because my daddy started us out at an early age.

While the entirety of this quote is illuminating, one point stands out as especially important: the fact that Samuel and his siblings might have gotten in the way while their father was working on the farm. This signals that, even if the children were impeding work, their fa- ther knew the importance of teaching his children the intricacies and arts of farming. He wanted them there. Education was more impor- tant than efficiency. Samuel had a good and patient agricultural men- tor, which has served him and his farm well.

Beyond learning to do the work of stewardship early on, Samuel also gained familiarity with the place itself. Speaking to his enduring imagination, he said, "I'm pretty well acquainted with [this land]. Because when we were young boys and stuff, our daddy bought each one of us a pony. And we would get out here and ride at night. I mean, [it was] black dark, and [we'd] go all over the place. And to think back now," he continued, "when I get out and the cows won't come up or something, I get to walking in these hills and stuff. . . . I know my land. I can go in the dark anywhere I want because I was raised here. I know it, and I love it."

Affection

Like imagination, affection is consistently practiced over time. Usually substituting the word "love" for affection, the farmers I interviewed spoke to this virtue's prevalence, identifying it as a key stewardship motivation.

Walter's comments offer a helpful starting point for exploring affection. Speaking about the land that has helped sustain his family for over 150 years, he said, "I love that place. . . . I'm attached to it, and I love the property, and it's a proud feeling. I'm proud." His love and pride are certainly directed toward the farm itself. But subsequent comments also emphasize that his affection is directed toward family members, past and present, who have tended and been supported by the farm. These are people, Walter explained, who have had a profound impact on the world. They participated in sit-ins to desegregate public spaces, helped integrate Maury County schools, worked multiple jobs to send children to college, and farmed successfully for generations. As Walter stressed with his remarks and memories, people are an integral element to place. To illustrate his affection for his family's farm, Walter shared one of his most meaningful memories:

It wasn't a fun time. It was a real sad time. . . . It's still something I think about. My grandmother died on Thursday, [*specific date omitted*]. She had been sick in the hospital that week. And I want to say that Wednesday, we had gone out and cut a field of tobacco, and it was still sitting in the field. That Saturday morning, before they had her funeral, her grandsons and her nephews and her son went to that tobacco field and got that tobacco up out of that field and put it in the tobacco barn. Now, we didn't hang it, but we got it up out of the field before rain could get it. And that right there is something that stands out to this day. . . . And I thought that the fact that everybody came and everybody got in there and worked— something that probably would've took all day took about three hours. We got it up, and then we went and laid her to rest.

Walter and his family's tobacco-patch efforts were an ode of affection. In caring for the farm, they showed their united love for the place— which included, even in death, Walter's grandmother. After finishing their work, the family could grieve.

Ronald, like his brother, said that family has enhanced his affection for place too. He recalled how stories told in the stripping room—a common way to pass the time while pulling thousands of tobacco leaves off stalks—deepened his love for family and farm. But as the family member who does most of the farming now and as someone who has been an active caretaker for decades, he also had plenty to say about affection for the land itself. After musing on his steadfast love for the farm, Ronald laughed and said, "I tell people all the time that if something ever happened to it, I don't think I could even come down that road anymore." When I asked him to elaborate on this comment, he said, "You've got to fall in love with [the farm] to appreciate it. I know it's a weird feeling, but it's just like at the beginning of hay season. After the first cutting, you smell the hay cure at night. It's different things like that. It kind of feeds you. It fuels you, and it keeps the fire going. Seeing that first roll of hay come out [of the baler] and knowing that the cows are going to have plenty of hay to eat for the year . . . It's kind of a weird feeling, you know?"

To briefly insert myself here, I'll answer Ronald's rhetorical question: yes, I do know. There is an almost unexplainable spiritual contentment in knowing that animals in your care will be happy and
healthy during difficult winter months. In early 2021, for example, I
spent a morning scooping manure and bedding the barn on my family's farm. Those who've had similar farming experiences know that
this is tough and itchy work. Loose pieces of straw worked their way
up my sleeves and down my back, and the multiple layers I wore to
keep warm did their job too well, causing me to sweat. Before long, I
shed my insulated flannel shirt and knit cap, leaving them hanging
over the wooden feed-room gate, the same one my father and I had
rebuilt and hung years before. Now down to a long-sleeved shirt, I
could see steam rising off my shoulders. With the barn bedded, I
climbed into the loft and threw down fresh bales of hay—hay that my
father, mother, brother, wife, and I had cut, raked, baled, and hauled
the previous summer—into a wooden trough that has been in our old
barn for generations. As I worked, cows and calves streamed in, taking
shelter from the cold and wind. They ate and then settled into freshly
strewn straw. Despite the weather, they were dry, warm, and full. In
that moment—entirely alone, save for the company of cattle—I felt
more peace and purpose than I have in years.[1]

But back to Ronald. Stoking his affection through sight and smell
keeps him going, even when the going gets tough. As others have
done, he was quick to mention that farming can be difficult and
sometimes frustrating work. But after these remarks, he reemphasized
how affection drives his stewardship. "You either love [farming] or
you don't. You've got to have passion for it. You've *got* to have a passion for it. Seeing those cows eat in the wintertime or the horses or
whatever, . . . it's all worth it in the end."[2]

George echoed that passion. He learned to love the land from his
parents and older siblings. As he points out in the following quote,

this love did not go untested. Despite challenges—what he calls the "minuses" and "mistakes" of farming—his affection has both endured and evolved. Answering a slight variation of Wendell Berry's question "Why do y'all continue to farm? Why do y'all continue to care for and own that place?" George said,

> I always looked at it as I have had the advantage of seeing everything by being the small guy who grew up behind everybody else. I got to see all the pluses and the minuses, the mistakes and the great accomplishments. And there have been more pluses than minuses and more accomplishments. It's something I just grew to fall in love with. And that's why now, I can go there and do anything, and it just feels good. It's hard to explain. And it's something that, in talking with someone [without the same experiences], they won't understand. It's just foreign to them. But it's in me, is what it is. It's in me. It's there.

George's insistence that his affection is ingrained—"It's in me. It's there."—reinforces the idea that this virtue is core to his character. Though other people may struggle to understand his perspective, George's affection for the land is ever present.

Ruth, too, described her affection in the present tense: "I love the land—that is so true." However, a story from decades ago perhaps illustrates her affection best. In the late 1970s, Ruth and George's extended family members sold some of their farm. The land was located adjacent to the farm that Ruth and George grew up on and now co-own with their siblings. Determined to purchase what land she could to keep her family's farming legacy going strong, Ruth resolved to go to the auction. In preparation for the sale, she told her husband, whom she calls "a city guy," "We're going to buy some land next to my parents." Rather than being wary of Ruth's statement, her husband embraced it. "It was not, 'I don't want to do that,' " she said about her husband's response. "Because he knew how much I loved that land."

When auction day finally came, Ruth and her husband drove out to the farm. The acreage they were eyeing came up for sale, and they got in a bidding contest with another man who wanted the same land. Slowly, the price climbed until it more than doubled what the other tracts of land had sold for that day. But that made little difference to Ruth. Sensing her determination, the competitor gave in. At the end of the day, Ruth and her husband added eleven acres to their family's farm, land that—because of its reliable water source—has proven essential for the family. When faced with financial sacrifice, it was affection that powered Ruth's resolve.

Like other participants, Samuel expressed a deep and enduring affection for the land, repeatedly saying "I love it" throughout our conversation. And while his love for the place extends back into childhood, it has expanded even more, at least geographically, in recent years. Samuel explained that when his parents passed away, he and his siblings decided to transfer the family farm to their youngest sibling. Handling farm ownership in this way would help prevent heirs' property problems in the future, giving the land a better chance at remaining in agricultural use and in the family. A little over a decade ago, Samuel and his wife were able to purchase a farm that borders the original family place on three sides.

Buying this land—which had also been owned by Black farmers and was part of the same area Samuel and his brothers explored by ponies as children—was literally a dream come true for Samuel. "I always dreamed to own a farm," he said proudly. "I finally got able to buy a farm where my homeplace is at."[3] Importantly, the farm has also become a living dream for Samuel's wife, despite her initial reservations. "To make a long story short," Samuel started, "I married a girl from Columbia, and she ain't never been on a farm." A few months after moving out to the land, Samuel asked his wife if she was happy. Laughing, Samuel quoted her response. "She said, 'There ain't no way you could make me move back to [town].' So I love it," he continued, "she loves it, and my kids love it."

The fact that Samuel's children have affection for the farm is significant, and it's not accidental. Just as his father helped him get to know the family farm through shared farm tasks and pony rides, Samuel has encouraged his children to spend time on the land. Through working and playing there, they have grown to love this place. Samuel explained that his children, now adults, have always been eager to help out on the farm, fixing fences and working livestock. They enjoy relaxing on the farm too and especially love to camp out by the pond and go fishing. With their parents' help, Samuel's children are passing this affection for place down to future generations. After Samuel explained his infectious love of animals, he said, "I've got grandsons and granddaughters—I've got little great-grandkids now—and they think the world when we take them down there to feed the goats and mess around on the farm." In addition to affection's intrinsic value, Samuel mentioned an instrumental reason that he is working to help children, grandchildren, and great-grandchildren develop their own love for the farm: "I hope they take it over. That's what I'm trying to get them into, so that the farm won't be sold off in pieces." Other farmers described trying to be exemplars of affection too, in the hopes that fidelity follows.

Fidelity

Imagination and affection give rise to an ultimate stewardship virtue: fidelity. Defined by a commitment to care for a place even in the face of difficulty, fidelity is a hard-earned but joyful virtue that helps people endure through adversities or challenges. Refined and enhanced through continuous practice, it prioritizes resilience and embraces loyalty. Eventually, it becomes a central element of a person's character. In similar and unique ways, farmers described their fidelity to place. In doing so, they attest to the on-the-ground impact of this virtue.

Through several different comments, Ronald showed how fidelity has directed his caretaking efforts. A part-time farmer—he works a full-time job as a maintenance technician in a factory, in addition to coaching youth sports teams for his children—he carries out his stewardship responsibilities whenever he can find time. On occasion, that means taking hard-earned vacation days, time meant to help him rest and recover from a physically demanding job, to care for the farm. "Last week, I took vacation," he mentioned during our conversation, "and I was up at six o'clock every morning and didn't go to bed until ten thirty at night. A lot of the time, I wasn't coming out of the field until dark." These "days off" were long. At the time, the weather was brutal, with high temperatures regularly reaching the upper nineties. But Ronald embraced this work, knowing that his efforts would lead to a successful hay crop. "You've got to be able to put the work in," he explained. Acknowledging the multiple challenges he and other farmers—especially Black farmers—endure, he said, "You've got to know it's going to be a struggle and that you're going to be in an uphill fight."

For Ronald, this "uphill fight" is about more than just providing fodder for livestock, repairing fences, and mowing fields. It's also about caring for home.[4] On this notion of home, Ronald shared a story about him and his wife looking to purchase a new farm. "We contemplated maybe selling [our land] and moving to a different location. We looked at buying a farm a couple years ago, but every time we went over there, it never did feel like home," he explained. "I know it takes time to get used to things sometimes. But also, you know when it's home. Over the years—from the time you grow up playing on it and doing different things—you can feel that feeling of home.

Do you know what I mean?" Stemming from his imagination and affection, Ronald's fidelity to place has kept his family rooted where they are. Recognizing his responsibility as a steward of homeland, Ronald concluded, "I want to continue to keep this farm as long as we can."

Ronald's brother, Walter, echoed some of these comments. For these siblings, the family farm carries meaning that transcends agricultural production. But alongside Walter's fidelity to place lies an emotion: fear. When I asked him about the future of the land, Walter responded that he is nervous. "See, that's a scary thing to think about," he admitted, making specific mention of his family's long history on the farm. Anxious that something will happen to the land—either through heirs' property issues or generational lack of interest in agriculture—he said, "It's kind of scary to think about the future. And we are some of the last remaining Black farmers to begin with, you know? In a generation, the land might be gone. And that's something I don't think anybody ever thought about happening." In reference to his fear for the future, he also noted the financial difficulty of simply hanging on to the property. For him, his brother, and other family members interested in keeping the farm, ownership can be an economic challenge, "especially with $600,000 houses popping up around [the farm]" that increase their property taxes and change the local landscape.

Like Walter, others mentioned fearing for their farms' future. Given the situation, it's an understandable emotion. But more emphasis was placed on shared stewardship commitments. Ruth, for example, spoke at length about her fidelity to the farm. "My youngest brother [George] and me—the oldest sister—have a mind-set that we are not ever going to sell," even though doing so would lead to an economic windfall. Other siblings do not share those sentiments, she said, but George and Ruth hold their ground. "We just kind of agree to stay consistent with 'We're keeping the farm,' " she explained in reference to land-related conversations with siblings.

Preferring to use the word "responsibility," Ruth continued sharing her thoughts on fidelity to place. Explaining that this virtue has come to influence all aspects of her life, she said, "It's important that ownership has responsibilities." In other words, fidelity means more than just keeping the place. It also means providing good, thoughtful, and consistent care, even when it's hard.[5] Only then can one be a great steward. Ruth has tried to instill this sense of responsibility in her children, and she noted that—at least with regard to the farm—her daughter has begun to embrace it. Her daughter learned this life lesson partly from raising a newborn calf into a full-grown cow as a child. This work, which required patience, was done alongside her mother and her grandparents. Because of this experience and others, Ruth said her daughter now fiercely defends the farm: "If anybody says anything about selling [the land], she goes into a hysteric fit. 'We're not selling anything! We're not selling anything!' That's what she always says." While her own work as a caretaker is not yet complete, Ruth has tried to be a role model of responsibility and fidelity for future generations.

George, too, has aimed to instill these lessons in his children. Thanks to his and Ruth's efforts, authentic hope exists for their family's continued care of the land. But in the present—and for the foreseeable future—George plans to continue tending the farm himself. For him, practicing fidelity is a pleasure. When I asked him how much longer he plans to continue farming, he immediately replied, "I'm in it until they put me in the ground." Like his sister, George knows they could call a real estate agent, advertise their land in the newspaper and online, and quickly make lots of money. But his reference to a lifetime of welcome labor alludes to Berry's marriage-focused discussion of fidelity. As in a union between two people, George's commitment to the farm is clear. In this instance, *'til death do us part* defines his relationship to the land.

Samuel mentioned a lifelong desire to keep stewarding his land as well. As quoted at the beginning of chapter 5, he said, "I hope to die here on my farm. I ain't no way soon ready to go, but if I do, I hope I leave here on my farm. . . . I love my farm. I wouldn't take nothing for it." If he does die on the farm, Samuel will follow in the footsteps of his father. "[When] he was seventy-nine years old, he made me promise that I never would put him in a nursing home," Samuel shared. "He made me promise him. And things got rough, but I kept my promise, and he passed away in my living room looking down there at the lake." In Samuel's loving care, his father died right next to the homeplace.

This promise is not the only fidelity-driven commitment Samuel described. When he graduated high school in the late 1970s, Samuel made a promise to himself. In his yearbook, he inscribed a few paragraphs, laying out life goals he swore to achieve. "I had in my dreams that I was going to own a farm," he said, reflecting on this moment and reiterating an earlier comment. Given the difficulties that he knew were ahead, he had good reason to wonder if he could keep this vow. Sitting with family on his farm decades later, Samuel showed his hand-scrawled yearbook notes to his children. They were awed by his determination and success.

Even though people are actively offering to buy the land, coveting the beautiful panoramic views from the farm's hilltops, Samuel keeps this promise still. But like Ruth, he acknowledged that fidelity cannot rest in ownership alone: "It's my responsibility to care for [the place] and to keep it in good shape." He hopes that, through his example, future generations will make a comparable commitment to the farm. And he has good reason to believe that they will. Before we ended our conversation so he could get back to work repairing a busted fuel line on a piece of machinery, Samuel offered a closing thought, one he hopes his children, grandchildren, and great-grandchildren will remember. The land, he said, "is something you can't replace."

Stories from the past, actions in the present, and hopes for the future reveal that fidelity is essential to stewardship. Tending the land isn't easy. It inherently invites difficulty, especially given the additional challenges that Black farmers have had to navigate. But for the farmers featured here, fidelity remains an active virtue. It has motivated them to persist in place.

Other Stewardship Motivations

Imagination, affection, and fidelity are indispensable stewardship motivators. Still, the people I interviewed cited other factors that spur and sustain their care too. Like folks in Robertson County, farmers here mentioned economics, spirituality, and family as stewardship influences. They also spoke about pride as rural farmers and the importance of land through the lens of race. All these factors—particularly the latter two—are important on their own. They also supplement and strengthen stewardship virtues.

Ronald, as mentioned earlier, noted that it is difficult for smaller-scale farmers in Maury County to make money from agriculture. But he still finds ways to earn income from his farm. While it's not his primary agricultural motive, Ronald explained that turning a profit from selling cattle is very important to him, and he admitted that if he were to consistently lose money on the farm, it would be hard to keep farming. Walter made similar remarks. Building on prior comments about financial difficulties, he said, "The tax alone on [the land] is just frightening to think about paying. You got to make a profit on that land to turn over the tax on it."

Samuel, too, mentioned taxes and economics in our interview. He relies on his goats to help cover annual property tax expenses. In fact, he has adapted the goats' breeding schedule to fit this purpose. The mothers have kids, and "by the time tax season is ready, [the kids] are ready to take to market, and they take care of [the taxes]." "And if I ever get my cows set up right," he continued, "they'll take care of this other stuff that I need for the farm." None of the participants mentioned aspirations of getting rich through farming. But sustainable income—which helps support their stewardship virtues—was significant, and it again emphasizes that economic viability is critical for conservation.

Religion and spirituality also influence the stewardship of some farmers, though the impact is not decisive. Walter said that his Christian faith is strong, but it does not really play a role in how or why he cares for the farm. His brother, Ronald, said that Christian beliefs spur his care "in some ways": "You want to try to protect the land the best you can." But for him, faith and farming were more important to his grandparents and other older family members. Because of his full-time job, his familial and community commitments, and his general busy schedule, Ronald often finds himself farming on Sundays. "You make hay when you can make hay," he said, citing an old farming adage about taking advantage of opportunities. Older family members would have been in church instead of working, Ronald said. If they were still alive and knew that he or his brother was out building fences, planting gardens, or working cattle on the Sabbath, "they'd chop our heads off," he laughed.

Of all interviewees, Samuel expressed that religious faith was most important. He grew up in the Christian Church, and it has been a major source of inspiration for him. He has tried to pass that on to his children. Talking about his specific rural church and the sense of community it nurtures, he explained that his faith is "part of what's keeping [him] going, because everybody has some up and down days, but when you have a hard time, you can call on Him [God], and He'll never let you down." It's clear that his fidelity benefits from the comfort and resolve offered by his trust in God. When I asked him specifically if his faith impacts his stewardship, Samuel offered a moving response. "I think so. Because if it wasn't for Him, I wouldn't have none of this. He's my sole provider, and . . . How do they say it? 'This land ain't mine. I'm just a caretaker for Him.' " Sharing the same sentiments as some farmers in Robertson County, Samuel bolsters his care with his faith.

Among all participants, family was a commonly referenced motivation for caretaking. In addition to the earlier comments Ronald made about the importance of having farm exemplars in his family and hoping that the next generation will carry forward an agrarian tradition, he said, "If you've got pride in what your family has established and done for years, then for me—in my view of it—I want to continue to be what the family has established and show pride in what I'm doing, to continue to operate and do the best job I can, to make the family look good." Walter, Samuel, and George made comparable comments.

Ruth's references to family as a stewardship motivation are most noteworthy. She is particularly grateful for her grandmother, who refused to sell any of her land and chose instead to pass it down to the next generation. Partly because of her gender and partly because of her love of agriculture, Ruth mentioned feeling a special connection with this older woman. And although she acknowledged that some siblings do not share these feelings, Ruth views the farm as an essential part of her and her family's identity. When she was growing up, "everybody knew" her family as "the country bumpkins": "They knew we were farmers." Even today, she still sees herself and her family in that light. Although they also have professional, nonfarm jobs, "we're still farmers," Ruth said proudly. "I tell people, 'I am a *farmer*.' I own grass and dirt on [omitted] Road."

In the future, Ruth wants this agricultural identity to continue. When I asked what her primary motivation for keeping the family farm is, she replied, "I can tell you that mine is leaving a legacy. . . . That farming part of me is bigger than me being a computer person. It's bigger than me being [a community leader]. I want to leave a legacy for the next three or four generations that we were farmers."

This deep, durable pride in being known as farmers—as committed, constant members of rural communities—is telling, and it was reflected by others too. Folks understood that, as Black farmers, they are a small population. Outsiders—and locals too—sometimes look at them as if they are out of place, said George. He imagines people, surprised by their presence, asking, "Why do they do that? Why do they have that [land]?" But participants were insistent. In Maury County and beyond, Black farmers belong.

This truth of Black people belonging in and embracing the rural US as careful, capable, and fulfilled stewards of place is powerful.[6] In the moving book *Belonging: A Culture of Place*, the writer, feminist, and cultural critic bell hooks explores the idea that although the South has caused pain for many Black Americans, this land is still a home. It can provide liberation for those who love it. Or, as hooks stated in a recorded conversation with Wendell Berry, "Most people imagine that black folks working the land were just victims, working for little and living a starvation life. We both know that the life of a small farmer can be terribly hard. What outsiders rarely see is the spiritual reward—the power of redemptive suffering. . . . Much of what our nation has lost is that awareness that the earth can be for us a place of spiritual renewal. . . . It is a place where we can be transformed."[7] The farmers who shared their stories with me understand this fulfillment well.

Elsewhere, hooks—who calls Berry's vision of homeplace "a guiding light"—speaks further about the human-land relationships Black people have nurtured. In a chapter titled "Earthbound," she writes,

That sense of interbeing was once intimately understood by black
folks in the agrarian South. Nowadays it is only those who maintain
our bonds to the land, to nature, who keep our vows of living in
harmony with the environment, who draw spiritual strength from
nature. Reveling in nature's bounty and beauty has been one of the
ways enlightened . . . people in small towns all around our nations
stay in touch with their essential goodness even as forces of evil, in
the form of corrupt capitalism and hedonistic consumerism, work
daily to strip them of their ties with nature.[8]

Hooks wrote these words, and she lived them too. She was, and even
in death still is, an exemplar for others who feel called to live in and
care for rural communities. In doing so, she—like Berry—rejected
common metrics of literary and academic success. Years ago, she left
New York City and returned to the hills of Kentucky, her childhood
and ancestral home. In a rural community there, she "could be in
community with other folk seeking to revive and renew our local environment, *seeking to have fidelity to a place.*"[9] In many ways, the "culture of place" that hooks describes encompasses the purpose and pride
participants feel in farming, as well as the stewardship virtues that
keep them rooted in and caring for the land.

In *The Hidden Wound*, written in 1968 and first published in 1970,
Wendell Berry grappled with racism in the United States. Drawing on
the words of Malcolm X, Confucius, and others—and reflecting on
his own personal experiences of growing up in the segregated rural
South—Berry likened racism to an open, physical wound. For too
long, many white Americans have tried to ignore the nation's collective wound or treat it with a simple and superficial bandage. This has
led to disastrous results. In order to better move toward racial justice,
we must all acknowledge and try to understand the wound that is racism, Berry writes. We must cultivate authentic empathy as best we

can. Only then will the wound, which affects all Americans, stop fes-
tering and causing pain. Only then can it heal.[10]

This wound metaphor certainly applies to agriculture. Throughout US history, Black people have faced difficulty and discrimination
in farming. Racism has led to slavery, violence, land theft and dispossession, diminished educational opportunities, loan denials, and
more, drastically reducing the number of Black farmers and the
amount of land in their care. Some of these same oppressive outcomes
have also impacted other non-Black farmers of color, as well as many
Indigenous communities and poor people of all backgrounds.
Through firsthand accounts, farmers here show that less overt racial
injustices—such as verbal microaggressions that question a person's
presence and identity—have been damaging too. In farming, the
wound is deep for everyone involved, even though some feel it far
more acutely than others.

On top of racism, smaller-scale Black farmers in places like Maury
County have also been grappling with the simultaneous rise of farmland loss and agricultural consolidation. These side-by-side problems
have taken emotional, economic, and environmental tolls and generally made survival on the land more difficult. Over the past few decades, these farmers have watched their rural communities and
landscapes be reshaped in cultural, social, and physical ways.

In line with Berry's charge in *The Hidden Wound*, we must acknowledge all these struggles and sores. Doing so is honest and necessary work. But we must also recognize the people who, in the face of
adversity, have persisted in place. Against the odds, the Black farmers
featured here have continued to tend their farms, embracing their
callings and commitments to care for the land. They have felt these
wounds, gritted their teeth, and forged ahead, and they've brought
others along with them. Imagination, affection, and fidelity—as
well as other factors, such as a shared sense of belonging and a strong

"culture of place"—have been pivotal to their perseverance. Their stories and experiences highlight the importance of human-land relationships and devoted care.

We would be wise to emulate these Black farmers' examples and efforts. In doing so, we might begin to heal some of our nation's wounds: wounds of racism, environmental exploitation, economic injustice, and agricultural excess. We may find ways to better serve and love one another and to care for the places we call home. By generously sharing their experiences, these farmers help us see a better way to interact with the earth. They show us a way forward.

7

Leveraging Love for the Land:
From Philosophy to Practice to Policy

The farmers featured in this book prove the power of stewardship virtues. Ample opportunities exist for them to abandon their land, to sell their farms, to pursue easier lives. But as their experiences and stories reveal, they have maintained enduring commitments to place. As they navigate flawed systems that threaten their existence, they have waged what Ronald Jager calls an agrarian resistance of sorts, a "struggle for the soul of agriculture."[1] On the page and in the pasture, imagination, affection, and fidelity have thrived.

Farmers and nonfarmers alike can learn from their example—and should do so in a hurry. On global, national, and local levels, we face a series of crises. From species extinctions and environmental injustice to pollution and worsening climate change, not to mention the interconnected issues of farmland loss and consolidated industrial agriculture, our planet and the places that compose it are in urgent need of better care. Time and again, smaller-scale farmers in two rural Tennessee communities reveal that devoted stewardship depends on people's willingness to understand and commit to specific places on earth. With effort, others can emulate their example. By lovingly tending our own communities, we can serve our own homes while also becoming better, more empathetic stewards and citizens of the world.

Thinking about how these lessons apply beyond rural communities is important. No matter the geographic setting, people should embrace stewardship virtues. But reflecting on how human-place connections and character-driven care can be harnessed for greater good in places like Robertson and Maury Counties is essential too. By supplementing and strengthening strong personal commitments to place—by treating virtues as assets rather than liabilities—federal, state, and local policies can serve small and midsized farmers and their surrounding communities. This book concludes by asking: How can we leverage love for the land?

In these final pages, I offer a few suggestions for addressing the main structural challenges highlighted in this work: farmland loss from development, agricultural consolidation, and racial inequities in agriculture. Recognizing the immense value of grounded experience and knowledge, I also cite reflections and suggestions from farmers, farm-service providers, land conservationists, and community leaders.

I explore potential solutions only briefly. The primary purpose of this book is not to analyze policies or propose legislation—its goal is to reveal the power and potential of imagination, affection, and fidelity. Plus, hundreds of books, reports, and articles have already been written on farm and rural policy, and many organizations and programs are dedicated to good agricultural advocacy. The scholars and practitioners behind these publications and efforts speak more effectively to policy issues than I can here. Further, I recognize that the suggestions made here are but a small piece of the puzzle. Our agricultural system is, after all, a *system*, and one book chapter won't change it. Even with all these caveats in mind, a succinct overview of strategies to serve smaller-scale farmers and the land itself will point to tangible ways to secure a better future.

I should note the role of stewardship virtues in the strategies proposed. These proposals all depend on enduring commitments to place in order to succeed, and they flow from the ethnographic research at the heart of this project. Without virtues to guide on-the-ground efforts, the benefits of policy change will be muted. These strategies are not going to make small and midsized farmers rich, and they won't make the physical work of farming and stewardship any easier. But they could help conserve the foundation of agriculture: land. They could level the playing field for smaller-scale farmers and farmers of color, granting authentic opportunity in spaces where it has long been denied. They could curtail exploitation and advance economic justice in rural communities. They could enable established small and midsized farmers to better nurture virtues and tend their farms, and they could give aspiring and beginning farmers a fighting chance to live their dreams and love the land.[2]

Imagination, affection, and fidelity will be necessary for any of the following policies to have the transformative impact we need.

On the federal level, several actions could advance farmland conservation. Perhaps unsurprisingly, one of the most helpful strategies for limiting farmland loss from development is funding land-protection programs. The 2018 Farm Bill authorized $450 million each year for the Agricultural Conservation Easement Program (ACEP). This program—which purchases land development rights from eligible farmers and wetlands owners, giving them financial support while limiting or prohibiting development on their land in perpetuity—is essential for spurring farmland conservation in communities across the nation, especially those with weak or nonexistent state and local land-protection programs.

Congress also simplified the ACEP land-protection process in 2018. For example, the Farm Bill altered ACEP's previous requirements for "matching funds," the limited availability of which often stopped projects—particularly in less-wealthy communities and states—in their tracks. Now farmers can sell their development rights at half value through ACEP and donate the remainder of the easement's financial worth. This adjustment should facilitate more farmland protection, particularly in places where productive soils and other important natural resources are present.[3]

As it stands, ACEP is a helpful program that provides many agricultural, ecological, climate, and community benefits. But it needs to be improved and enhanced, especially considering that the 2018 Farm Bill actually allocated far less money per year to federal land-protection programs than the 2008 iteration. Plus, only a fraction of the $450 million allocated actually goes directly to farmland protection. The continuation of the program should be about more than conservation and environmental benefit. It must also prioritize and advance economic justice. More funding not only would conserve more land but would also enable smaller-scale farmers to better participate. New resources could be directed to support working-class farmers who want to protect land from development but need financial compensation to do so. Even receiving the aforementioned half value for development rights would be a boon to many farmers that could help them ensure a viable future for their farms and families. That means increasing the program's capacity and ensuring that it remains funded at a high level into the future. For a host of reasons, and especially those related to climate and the environment, write Peter Lehner and Nathan Rosenberg, "Congress should substantially expand ACEP."[4]

In addition to increasing funding capacity, ACEP and other land-conservation programs—like the Regional Conservation Partnership Program, a partnership effort intended to advance conservation across targeted areas—could be further streamlined. These programs are still exceptionally onerous for farmers and support organizations to navigate. It can take years and mountains of forms to use these programs for permanent land protection, a daunting process that often scares away or frustrates would-be conservationists. This difficulty affects traditional easement transactions, as well as the new "Buy-Protect-Sell" element of ACEP, which was intended to help state and local conservation organizations and agencies purchase land at risk of development, protect it via a permanent agreement, and then sell the land to a farmer. This program could be especially effective if protected land is sold at an agriculturally appropriate price to new-generation and underserved farmers, groups that struggle mightily with land access, security, and tenure. Yet the long and difficult process currently impedes equitable participation, slowing or stopping the disbursal of funds to people who could use the money to save their farms.

All of this is not to say that accountability or oversight should be sacrificed. Land-protection projects must be done well, especially since it is the public's money supporting ACEP. That requires checks and controls. But the bureaucracy of funded farmland protection could be reduced, leading to more on-the-ground conservation and a more efficient and effective use of public dollars. We have many brilliant people, including farmers, in the United States who design processes and programs that are both rigorous and functional. This same thoughtfulness can be applied to improve our federal land-conservation efforts.

The federal government should also concentrate on better supporting small and midsized farms. In *Family Farming: A New Economic Vision*, Marty Strange shows that agricultural policy has been favoring larger and larger farms for decades. Part of this bias toward big farmers stems from disdain for smaller farms shared by many powerful lobbyists, agribusiness specialists, and government leaders. Some experts, Strange writes, "have been known to get downright hostile about small farms. They berate them for being irrational, inefficient, a burden to society, selfish hoarders of land and other resources that bigger farmers could use better." Elsewhere, Strange uses data and qualitative observations to show these negative assertions are simply not true. In many cases, they are politically motivated. John Ikerd—a well-known agricultural economist and author—has also persuasively argued in favor of smaller farms, citing influential evidence that "small farms can be farmed sustainably, thus benefitting farm families, rural communities, the natural environment, and society in general." In *Farming for Us All: Practical Agriculture and the Cultivation of Sustainability*—an intimate examination of farming in Iowa—Michael Bell similarly shows that smaller-scale farms can be widely beneficial too, despite common misperceptions of and disrespect for these operations.[5]

I witnessed this sort of derision for smaller farms firsthand while doing ethnographic research in Middle Tennessee. Early in my research, I called a local farm-service provider's office to request an interview. I patiently and politely explained my project. Then I was scolded. After rebuking me and my work over the phone, a leading public agricultural service provider in Robertson County stated that small-scale farming was futile and that it was right for massive farms to take over. To illustrate his point, he argued that Walmart is much better than locally owned grocery stores. Small stores are obsolete, he said, a waste of time and space. To him, the same logic applies to small farms and their inferiority to large operations. In my view, chain stores like Walmart can have their place, as can large farms in certain situations. But it is troubling to envision a world where local stores, shops, and farms are eradicated. In fact, it's more than troubling. Given the many ills of large-scale industrial agriculture and uncontrolled monopolies, it is dangerous and destructive. I should note that this service provider declined my invitation for an interview and hung up on me.

Decisive federal action is needed to curtail and reverse the nationwide trend toward industrial agriculture. In a representative report for the USDA's Economic Research Service, Jonathan McFadden and Robert Hoppe speak to Strange's findings about disproportionate federal benefits for large farms. They reveal that support for larger, more industrial farms has increased in recent years. This favoring of large farms and farmers is particularly evident in government funding programs. "Changes in the structure of agriculture have changed the distribution of income support over time," McFadden and Hoppe write. "Specifically, commodity program payments, some conservation program payments, and federal crop insurance indemnities have shifted to larger farms as U.S. agricultural production continues to consolidate. Since the operators of larger farms have higher household incomes than those of smaller farms, commodity program payments and support through federal crop insurance have also shifted to higher income households." In other words, larger, wealthier, industrial farm operators are getting far more financial assistance from the government than are smaller-scale, working-class farmers. Some research estimates that roughly 80 percent of federal subsidies go to farms with over $1 million in annual revenue. And it's not just that more overall money goes to the largest, wealthiest farmers, which could be rationalized. In programs where money is distributed by the acre, a bigger farm means bigger payments. Recent evidence suggests that the biggest farmers also receive much more money *per acre* from major subsidy and crop-insurance programs than do smaller, less wealthy farmers— sometimes three or four times as much per acre.[6]

This skewed distribution of dollars offers unfair competitive advantages to industrial farmer-managers and big corporations and makes survival more challenging for less wealthy farmers. Oftentimes, bigger operations are defended because of their "efficiency." It's true that industrial agriculture produces a massive amount of cheap food and gleans sizable profits. It helps feed the world. But this efficiency is propped up by preferential policy. In other words, these massive operations claim that they're more efficient, so we offer them greater assistance and design larger systems to fit their needs, making them appear even more efficient and making smaller operations seem less so. But what happens when the playing field is leveled? Or when subsidies are defined by stewardship and not just acreage? Multiple people in Robertson and Maury Counties bemoaned these economic advantages for big agricultural operations, arguing that the farmers who need the least support are in fact getting the most.

Over time, much of this subsidy money should be redirected to benefit small and midsized farmers. Conservation practice dollars should be dedicated to these farmers too. Research shows that programs like the Environmental Quality Incentives Program, part of the Farm Bill's conservation title, "increasingly support large-scale, industrial farms rather than their smaller, more disadvantaged counterparts." In fact, a large portion of the set-aside funds for livestock operations in this program supports concentrated animal feeding operations "at the expense of small to mid-sized sustainable operations." Increased funding for smaller-scale farmers—which would help level the playing field so that the largest farmers are not always given unfair economic advantages—would help them continue or improve their land stewardship and agricultural production. Some funds could also be directed toward farmers of all sizes who are already actively prioritizing conservation practices, people who are often called "early adopters." Benefits could come in the form of direct assistance, favorable loans, debt forgiveness, and farm enhancements. Many smaller-scale farmers in Robertson and Maury Counties struggle to break even each year despite raising good crops and livestock. The same is true for similar farmers in other counties, states, and regions. Even a marginal increase in financial support could mean the difference between keeping and losing a farm. With rerouted agricultural assistance—or even just capping the public money pipeline to the biggest and wealthiest industrial farmers, a tactic some economists say would reduce federal spending "without any adverse implications for 90 percent (and in some cases, more) of U.S. farms"—taxpayers would benefit too. They could feel more justified knowing that their contributions are aiding people who genuinely need help rather than boosting the income of already-wealthy individuals. Further, a large percentage of redirected dollars should be reserved for supporting marginalized farmers and farming communities.[7]

Toward the end of supporting farmers who have long been discriminated against and underserved, the federal government should enact legislation that explicitly recognizes the history of racism and dispossession in rural Black communities and addresses the needs of Black farmers in the present and future. Legislators are slowly making progress in this realm. In November 2020, Senators Cory Booker, Elizabeth Warren, and Kirsten Gillibrand introduced the Justice for Black Farmers Act in the US Senate. This legislation was intended to shine a bright light on racial injustices and provide access to land for current and new Black farmers, as well as to advance conservation and regenerative agriculture among diverse farmers. It was also supposed to reform the USDA's civil rights process and provide legal support for Black farmers who face challenges from heirs' property and other issues. The bill won the endorsement of people like Leah Penniman, who is the co-executive director and farm manager for Soul Fire Farm, an agricultural initiative in rural New York aimed at liberation on the land. "For generations," she said, "the cries of Black farmers dispossessed of land and livelihood have gone unnoticed. In my twenty-five years as a farmer and land justice worker, I never dared to imagine that such an elegant, fair, and courageous piece of legislation as the Justice for Black Farmers Act could be introduced. We now have the opportunity to correct decades of discrimination, preserve agricultural lands, and equip the next generation of farmers who will feed the nation."[8] As Penniman notes, this policy change could begin to address centuries of injustice, though it does have shortcomings. In its focus on Black farmers, the bill appears to neglect other underserved groups, such as Indigenous farmers, Latino and Latina farmers, Asian American farmers, and others.

For the foreseeable future, the Justice for Black Farmers Act and related, updated efforts may go no further than being introduced. This and comparable bills have proven difficult to pass or implement. Other attempts to advance racial equity in agriculture have also been met with resistance. For instance, a portion of the American Rescue Plan Act of 2021—intended to provide much-needed relief in light of the COVID-19 pandemic—allocated over $4 billion for assisting minority farmers, largely through debt forgiveness. Many Black farmers were eager to participate in this program, and some signed paperwork and submitted it to the USDA, formally accepting the offer of debt relief. Yet, as has happened so many times before, the rug was pulled out from under these Black farmers, and the support they were promised didn't appear. The debt relief was held up in court because white farmers and well-funded organizations challenged the assistance and called it discriminatory. On this blocked support, Dānia Davy of the Federation of Southern Cooperatives—an established organization devoted to serving Black farmers—said, "It has definitely caused a very significant panic and a lot of distress among our members." Having invested in their farms in the meantime, some Black farmers now owe even more money than before, putting them at greater risk of losing their land.[9]

Moving forward, lawmakers must continue to push for racial justice in agriculture. In doing so, they can find creative ways to propose, pass, and implement reforms that can yield positive change and survive being challenged in courts. One such way to accomplish this goal is to blend efforts to advance racial equity with broader support for all owners of small and midsized farms. National data from the USDA shows that farmers of color tend to operate on smaller acreages. In fact, most Black-operated farms—about 85 percent—had fewer than 180 acres in 2017.[10] Thus, efforts to support small and midsized farmers in general will also directly help a more diverse farming population and perhaps avoid claims of discrimination.

One way that the federal government can accomplish this dual goal—advancing racial justice alongside equity for all small farmers—

is to create an Office of Small Farms within the USDA. An idea proposed by American Farmland Trust and the Black Family Land Trust, the Office of Small Farms could "serve as a coordinating body, bringing together representation from USDA's various agencies to identify additional needs for small farmers. . . . It would have dedicated staff, some of which should be experienced with providing on-the-ground support to small farms and ranches and diverse populations." This office—if funded, staffed, and supported at meaningful levels, enabling it to do more than provide lip service to its constituency—could continually press the USDA, Natural Resources Conservation Service (NRCS), and the entire federal government to devote more resources to and design programs for smaller-scale, diverse farmers. Along with other measures, an Office of Small Farms could help to flip our current agricultural system to one that encourages right-sized farming, discourages overindustrialized agriculture, and benefits more people and places writ large. "If large farms and corporate entities remain consistently advantaged over small farms and businesses," write Kaitlyn Spangler, Emily K. Burchfield, and Britta Schumacher, "then alternative agricultural schemes will be limited."[11]

The measures just described are but a few of the ways the federal government could address problems in US agriculture. Other actions would be helpful too. The government could enforce antitrust laws more vigorously, curb large-scale concentrated livestock production, decentralize the meat-processing industry, and support local processors, all of which would give more options and autonomy to small and midsized farm operations. After all, concentrated animal feeding operations, or CAFOs, are notoriously awful for the environment, workers, adjacent communities, and livestock, and just a handful of companies control the majority (70–90 percent) of US meat processing. With better regulations and policies, that could change, helping lead to a more resilient and sustainable food system. The federal government could also create and enhance programs to promote climate-smart and localized farming, make crop-insurance enrollment contingent on participation in conservation practices, and consider longer-term farm bills that prioritize conservation and sustainability. All these initiatives and more would help the US agricultural system begin to move toward systems and scales that allow imagination, affection, and fidelity to prosper. While unpopular among the few who benefit from our current system, these ideas may appeal to the many. Research shows that rural people—small farmers and nonfarmers alike—are concerned about status quo agricultural and environmental issues and open to reform, despite how such people are often represented in the popular media and imaginary.[12]

State governments could also lead efforts to address farmland loss and racial injustice. While states may also be able to tackle agricultural consolidation through measures like increased property taxes on the largest of farmers—those whose sprawling operations resemble open-field factories more than agriculture and could be taxed as such—or enforcing their own antitrust laws, the impact may be limited. With regard to structural change, the federal government is better equipped to handle reforms that challenge consolidation. States could, however, enact programs and policies that better protect farmland from development and advance justice for Black farmers.

On racial equity, states could pass legislation to remedy problems with heirs' property, an issue that—as shown in earlier chapters—affects many rural Black landowners. In particular, proposing and enacting the Uniform Partition of Heirs' Property Act (UPHPA) would help stifle the dispossession of land from Black farmers and poor farmers of all backgrounds. Farmers like Ronald, who shared specific concerns about losing his family's land due to issues with tenancy in common, would benefit from this policy. The legal scholar Thomas Mitchell writes that the UPHPA, which has already been passed in

South Carolina, Alabama, Georgia, Mississippi, Arkansas, Virginia, and—as of spring 2022—Tennessee, among other states across the nation, contains three primary components. First, this act guarantees co-tenants the right to buy out any other co-tenant who petitions a court to force a sale. This measure alone would curb or eliminate many instances of developers or wealthier landowners buying one share of heirs' ownership and then purchasing the remainder of the property at a court-ordered auction. Second, in cases where the buy-out of a tenant does not remedy issues with tenancy in common, the act favors partition in kind over partition by sale. Effectively, this means that—rather than a co-tenant being forced to sell their interest in land—the property itself may be physically divided into propor-tionate shares to satisfy all co-tenants. Finally, in cases where heirs' property *does* need to be sold, the UPHPA favors an economically equitable sales procedure. Rather than public auctions, which often yield lower values for heirs' property, this legislation prioritizes open-market sales where a disinterested real estate agent aims to secure the highest possible value for a given property. Mitchell, who won a pres-tigious MacArthur Fellowship for his work on this issue, calls this set of changes "the most substantial reform of partition law in the United States in 150 years."[13]

In addition to enacting such legislation, states—and nongovern-mental organizations—should work to help interested heirs' property owners clear the title to their land, especially by providing affordable or no-cost direct legal assistance to families. The federal government could assist here too, offering grant funds rather than just loans to heirs' property farmers and landowners. The UPHPA is an excellent piece of legislation, and it would be wise to enact it in every state. But it can't be the end of the effort. Forced partition aside, heirs' property owners continue to face many problems that aren't resolved by the leg-islation, including tax foreclosure and the inability to secure mortgages because of cloudy title. They also face more difficulty accessing conservation programs, implementing long-term on-farm conservation practices, and installing new agricultural infrastructure. As we move forward, we must continue to work with and take direction from individuals and communities that have long led in this space, as well as other academic and nonprofit groups devoted to the cause.

With reference to protecting farmland from development, states should set aside meaningful funding for agricultural land protection. If nothing else, financial assistance should be made available to offset the transaction expenses of placing a conservation easement on a property. Several states—such as Pennsylvania, Vermont, and Maryland, national leaders on this front—have taken these sorts of steps already. Many others—especially in the South—have not.

Several farmers interviewed for this project, for example, expressed interest in protecting their land via easement, but they voiced concerns about covering the costs of doing so. Melissa, who works for a statewide land-conservation organization, explicitly mentioned this challenge too. "There are costs associated with doing an easement," she explained, such as land surveys, title work, legal fees, and more. "Some of those [costs] can be significant for a person who is lower income and works paycheck to paycheck and doesn't have a lot of savings. I think in Tennessee, unfortunately, we don't always have a lot of ways to help those people, whereas other states might be able to even purchase easements and pay those landowners for conservation easements. We don't quite have that [right now]." For farmland conservation via easement to become a legitimate option for people of all economic and geographic backgrounds, assistance is needed, especially for lower-income farmers whose land may not meet qualifications for the federal Agricultural Conservation Easement Program. As Melissa explained, more progress could be made in the realm of farm conservation if additional funds were made available to purchase easements, even if not at full value.

Yet another farm preparing to be sold via auction. Stronger policies and heightened conservation support could keep places like this in active agricultural production and help new and beginning farmers access land. (Photo by author)

Money to support working-class farmers who would like to conserve their land could come from several places. In Tennessee specifically, the state's Department of Agriculture could set aside funds from its budget for this purpose. The Tennessee Department of Environment and Conservation could do the same. The state could even seek to create new sources of funding for farmland conservation efforts, perhaps coordinating with counties to assess agricultural-impact fees levied on developments that convert farmland to nonagricultural use. Money generated from this fee could be used to pay for farmland conservation efforts. Although national surveys reveal a bipartisan willingness to enact new taxes to support land conservation, this approach may be a long shot in fiscally conservative states like Tennessee. Still, a successful 2012 campaign to extend state land-conservation funding in neighboring Alabama shows that an effort like this is possible with careful coordination and planning.[14] Speaking to the importance of aggressively confronting farmland loss with public dollars in Robertson County, Frances said, "It's going to take public money. If open space and farmland is important, it's going to take federal and state money to make that happen. We've got to feed the world, and so we can't develop all this land. We can't have it all filled up with houses. . . . It's in all of our best interests to see some federal money directed in that way and some state money directed in that way."

The state could support farmland conservation in other ways too. For example, Tennessee could require all counties to develop a farmland protection plan, one that sets goals for conservation, identifies the importance of rural communities, mitigates haphazard land development, and proactively envisions a future for local agriculture. Or if not require farmland protection plans, the state could at least encourage them and offer grants to counties that develop and adopt them. These sorts of plans have proven effective elsewhere, from North Carolina to California and communities in between.[15] The state could also address the issue of generational turnover, an issue that farmers, community leaders, and land conservationists all touched on in our conversations. One way to tackle this problem is to offer more support to Tennessee FarmLink. This program seeks to connect farmers who want to sell their land with farmers, especially new and beginning farmers, who are looking to purchase acreage. Tennessee FarmLink is currently managed through a partnership between the Appalachian Resource Conservation and Development Council and the Tennessee Department of Agriculture. The program's location in the northeastern corner of the state and its limited funds and staff make statewide success difficult. But if given a boost, the program could help aspiring farmers and exiting farmers—particularly those without heirs or those whose children aren't interested in farming—while ensuring that land remains in agriculture. Research shows that farm-link programs elsewhere, such as in the Northeast, can be effective, especially when paired with wraparound services and individualized assistance. These programs can serve as trusted vehicles to provide in-depth estate and farm succession planning to new and retiring farmers alike. With the impending generational transfer of massive amounts of agricultural land, these services are crucial for protecting farmland and supporting farmers of all ages.[16]

New and revamped federal and state programs and policies would lead to positive changes in the agricultural and conservation spheres. But perhaps the most attainable changes could occur at smaller scales, where leaders are quite literally closer to the ground. Local-level initiatives would have a major impact on the well-being of small and midsized farmers, especially with regard to racial equity and farmland conservation. Here, we benefit from the perspectives of the farmers, farm-service providers, land conservationists, and community leaders whom I spoke with, who specifically reflected on the ways that strategic local efforts could serve rural communities.

Toward racial justice in agriculture, Black farmers mentioned that increased access to educational opportunities would be helpful. Classes, seminars, and trainings that cover financial topics—such as securing loans, accessing government grants, managing farm finances, and more—would be especially beneficial, explained Ronald. Walter believes that making programs like Future Farmers of America and 4-H more accessible to young people of color would be beneficial too, as would expanding access for youth to programs like Minorities in Agriculture, Natural Resources and Related Sciences (MANRRS). The goal in doing so is not to force students of color into taking farm classes or getting involved in agricultural extracurriculars. Instead, it is to ensure that they can pursue these options if they so choose and that stereotypes of nonwhite youth not belonging in agriculture are challenged and shattered.

Extension agents on the county level and agricultural teachers in public and private schools would benefit from further examining past and present racial injustices in agriculture, via both formal education and professional development. This might help them better understand, for example, why less than a dozen Black farmers remain in Robertson County and why Black farmers make up only 2.4 percent of producers in Maury County, a dismal stat that is nonetheless *higher* than the national figure. If in-depth training programs are not already offered, they should be designed and made available—or mandatory— for these extension and education professionals. While more of a state issue, it would also be valuable to increase the capacity of historically Black colleges and universities that train agricultural educators so more communities of color can learn from people who look like them. As in all aspects of life, representation matters in the agricultural education sphere.

Of particular importance, according to Black farmers I interviewed for this research, is farm-service providers and conservationists making an effort to engage on the ground with people of color. For example, the farm-finance classes mentioned earlier could be taken from extension offices and out into rural community centers, schools, civic buildings, or churches, if these venues are open to hosting. School advertisements for agricultural clubs could be sure to feature Black students in empowered and successful roles, and they could invite farmers of color and advocates for equity in agriculture to speak to their chapters. Youth agricultural groups could study the history of racial injustice and underrepresentation in agriculture, as one FFA chapter in a small, West Tennessee town has done in recent years.[17] They could deliberately encourage more opportunities for diverse membership. As Ronald mentions in chapter 5, working to authentically engage Black farmers and communities of color is essential to establishing trust, addressing systemic injustice, and creating a more inclusive agriculture.

In addition to advancing racial equity through education, local initiatives could help protect farmland from development. The farmers I spoke with expressed support for these kinds of policies and programs, as did land conservationists and farm-service providers. And according to multiple leaders in the Robertson County community specifically, some public officials would embrace these efforts too. Speaking of small-town mayors throughout the community and state, Kyle said that many elected leaders are eager to maintain farmland in their communities. Charlotte, who is one of these local leaders, affirmed this thought.

One way to conserve agricultural acreage in places like Robertson and Maury Counties is through stricter planning and zoning. To be fair, both Robertson and Maury Counties have developed comprehensive development and land-use plans. That's more than many counties across the South and nation can say. In these plans, maintaining rural character and conserving natural resources, like farmland, have been identified as priorities.[18] But as quantitative data and qualitative observations suggest, these priorities seem a bit hollow, as development has aggressively reshaped rural communities.

If local officials want to better address the erasure of agricultural lands, they might adopt larger minimum lot sizes—fifty acres instead of five, for example—in certain rural parts of their counties. They may prevent the expansion of larger sewage or water lines to limit sprawl, a strategy that Kyle mentioned in our interview. His rural community in northwestern Robertson County has not seen the same explosive growth as other communities throughout the county, partly because his area still has limited—yet adequate—infrastructure. Leaders should also think proactively about solar development, which is projected to cover millions of acres of rural land in the US by 2050 to meet decarbonization goals. Done well, solar development could serve rural communities, help boost some smaller-scale farmers' economic viability, and advance clean energy. But that's only if development is handled and regulated with extreme care—and with an explicit eye to farmland conservation and the well-being of rural places and people. Otherwise, it could become yet another exploitive, extractive industry.[19]

To continue the proactive planning theme, local leaders (with the help of stronger state support and more vigorous enabling legislation) could develop "agricultural districts," which David Bengston and colleagues describe as "legally recognized geographic areas designed to keep land in agricultural use." Enrollment in these districts could be made voluntary for farmers. In exchange for enrolling and limiting future division or development of land for a period of time, a farmer could receive economic incentives that help compensate them for the ecosystem benefits they provide through maintaining open space. They might be excluded from eminent domain considerations that could otherwise take parts of their property for public uses, such as schools or highways. They might also be made eligible or given higher priority for purchased conservation easement programs, either from the state or—if adopted on the local level—the county or town. Perhaps they could receive a greater property tax reduction than farms not enrolled in these districts. Neighboring North Carolina, where roughly 12,500 farms and 876,000 acres across ninety of the state's one hundred counties are enrolled in Voluntary Agricultural Districts, offers a workable model for this planning approach. Programs in Ohio and Virginia can also serve as useful examples.[20]

Logan, who is highly familiar with Robertson County's planning and zoning program, believes that more explicit, forward-thinking, landscape-scale protection of farmland and rural areas is necessary.

Agricultural districts are practical and could be very useful. These initiatives are "the biggest thing that the current [land-use] plan lacks. It looked at areas where there could be growth": "I think it needs to go a step further and protect some of that environmental and agricultural land, for environmental reasons and for hindrance-to-development reasons, to preserve that in perpetuity."

Officials could also raise local development-impact fees on land that is converted from agricultural to residential, commercial, or industrial use. Kyle explained that, in recent years, Robertson County slightly raised its development-impact fee in order to increase funds for public schools, angering some developers. About this increased fee, Kyle said, "Look, if you're going to come in here, you've got to pay. If you're going to play, you've got to pay. I still have that feeling. I'm not against growth, but it's got to be planned. You can't put the burden of growth on the people that are already here." Kyle said that Robertson County's development-impact fee, though recently raised, is not nearly as high as it is in other Tennessee counties near urban centers. For that reason, it seems that Robertson County—as well as many other rural communities and counties, like Maury, that are witnessing excessive farmland loss—could raise their impact fees, especially in areas zoned specifically for agricultural use. Officials could set aside the new revenue from these fees exclusively to fund land-conservation assistance programs or to purchase development rights from willing and qualifying farmers, accomplishing the dual purpose of slowing down development and providing fiscal support to farmers and rural communities.

John, who is a longtime town manager in Robertson County, is especially open to this idea of local government buying development rights from farmers. Although he and other leaders in his town have already "worked very, very hard" to preserve their farmland and to "try to prevent its loss" through measures like driveway fees and limited sewer expansion, John wants to do more.[21] In the recent past, he and others have explored grants that would enable their town to purchase development rights from farmers and protect farmland. Sadly, they have had trouble accessing those funds. "We are willing to do the work," he emphasized at one point in the interview. "All those things are steps that we've taken to try to preserve our farmland and our scenic beauty and rural character."

Rather than choosing just one of these farmland protection options, local leaders who are committed to conserving agricultural land should enact multiple policy changes. A suite of solutions is needed. As Tom Daniels writes, "The protection of a working landscape is a complex issue. No single protection technique can address all the dimensions. Several different techniques are often needed as part of a protection package to make a comprehensive yet flexible response to development pressures."[22]

There is no better time to work on comprehensive farmland protection than now. People in Robertson and Maury Counties—as well as other communities throughout the state, region, and nation—are witnessing firsthand the steady sacrifice of agricultural acreage. The issue is fresh in people's minds. Their emotions are raw. Renewed or heightened awareness, or what some policy scholars call "external shock," of farmland loss could engender enhanced public support for conservation efforts and stimulate effective, on-the-ground change.[23]

Amplifying efforts to protect farmland, enhance racial equity, and serve smaller-scale farmers at all levels would certainly benefit the people and places highlighted in this research. Evidence shows that these actions would serve others too. In a 1994 essay, Wes Jackson wrote that "the realities of [agricultural] industrialization are all around us. No 'ain't-it-awful' checklist is necessary."[24] He's right: examples abound and are easily found. Still, I've referenced some of these ills in other parts of the book and tried to explain why we need to change our current approach to enrich the collective good. I have also tried to show the need for advancing equity. Beyond what I've written, I hope the value of fostering a more just agricultural system is implicit. But because many people question or misunderstand the public benefit of farmland conservation, it's useful to recognize justifications for farmland protection here. While cultural, social, agricultural, and even psychological benefits exist, it is especially worth touching on the economic, environmental, and ecological reasons for farmland protection. Understanding and communicating these realms can establish momentum and broad public support for protecting agricultural land from real estate development.

When proposing a new subdivision or strip mall that will erase farmland, developers and proponents of sprawl claim that their efforts will be economically beneficial. In the short term, they're correct, especially if you focus on land values alone. Charles Francis and coauthors write that there are immediate economic benefits that accompany developing farmland for residential and commercial uses. "Factories, housing subdivisions, commercial malls, highways, and other intensive uses add immediate value to converted land. Yet," they continue, "the long-term consequences of losing agricultural production are not part of the accounting."[25]

One of these consequences is often a fiscal burden for local governments. Cost of Community Services (CCS) analyses explore these long-term economic effects. They aren't perfect, but these studies are a well-known way to estimate the impacts of varying land uses within a community, write Matthew Kotchen and Stacey Schulte.[26] CCS analyses divide land into different use classes and then compare the public expenditures associated with each land-use type to the tax revenue generated from these areas. Often, this results in a ratio that compares one dollar received to the amount of dollars spent to provide services. If an area generates more revenue than expenditures, it is deemed to add real value to the local community. Because CCS studies look beyond immediate impacts and instead describe how the development of farmland affects a local government's finances—and thus all citizens—over time, they highlight the financial upsides of farmland conservation in a way that developers' analyses do not. At a time when local governments are under unprecedented fiscal pressure, the long-term benefits of conserving farmland should be front of mind.

Consistent with a range of studies conducted across the nation, a CCS analysis conducted in Robertson County showed that farmland generated much more in tax revenue than it required in expenditures. Residential land uses, however, did not offer a positive economic return. Specifically, the study—which was performed by American Farmland Trust and funded by the Tennessee Advisory Committee for Intergovernmental Relations—found that for every dollar generated in tax revenue from agricultural lands, Robertson County only had to spend $0.26 to provide services to these areas. For residential land, the county had to spend $1.15 to provide services for every dollar of tax revenue.[27]

Whereas pastures and fields need little financial support from local governments, residential land uses require a multitude of services, including schools, sewage, emergency services, sidewalks, large water lines, roads, and more. "Because agriculture contributes to the local tax base without the costs attributed to urban development," writes Jeanne White, "it is an important asset for communities to preserve."[28] Agricultural land uses are pulling their own weight and then some when it comes to public finances in Robertson County. It is likely that this is the case in Maury County too, as well as in many other rural US communities. These types of studies should not be used to justify a total prohibition on development in these or other counties. People need places to live, work, shop, and eat. But they effectively challenge short-term economic thinking.

John offered a lived assessment of the theory behind CCS studies. In his work, he has observed that the long-term costs of providing for development are often forgotten or ignored by officials.

> That's something I think a lot of professional city managers don't seem to understand. In other words, when you recruit a subdivision, for example, . . . you get a fifty-house subdivision in your city limits, and people look at the property taxes that would generate. But the problem with that is the property taxes are not going to cover the expenses of providing services for that subdivision. It's really not an economic win to get all this growth that so many cities seem to seek. For a subdivision, you've got to provide fire protection, police protection; you've got to provide water and utilities and all those kinds of things, roads, stop signs, services, all those things. We absolutely save money by not having those [subdivisions].

His on-the-ground observations offer helpful insights into the economics of farmland conservation. Frances and Henry, both of whom are well acquainted with local land use economics, echoed the same sentiments.

Even beyond stable revenue to local government, farmland has clear financial upside. The Howard H. Baker Jr. Center for Public Policy at the University of Tennessee released a 2018 report highlighting the financial value of open space in a ten-county Middle Tennessee area. In the report's opening lines, the authors write, "Green open space—public parks, farmland, forests, wetlands—provide substantial economic, environmental, and health benefits. These benefits are lost when open space is converted to other purposes. Unfortunately, these benefits are poorly understood, leading to policy debates and land development decisions that ignore or undervalue these benefits."[29] The authors support these claims with quantitative data. While the report focuses on open space and not exclusively on farmland, evidence shows that—in Robertson County—open space contributes millions of dollars each year in community benefits: roughly $7.8 million via direct and indirect health benefits, over $5 million via outdoor recreation, and over $240 million in ecosystem services. Similarly, open space in Maury County offers about $9.5 million via health benefits, $7 million via outdoor recreation, and $362.2 million in ecosystem services. These numbers do not even include the significant economic impacts of agricultural jobs and localized economies.[30] All that to say: when farmland is lost, so is money.

Beyond economics, farmland serves important environmental and ecological functions, especially when compared to suburban sprawl. While farms, like any human activity, do produce an environmental impact—and can be quite destructive if certain practices are used—they are a healthier environmental alternative to sprawling, haphazard residential and commercial development. The environmental benefits of farmland are varied. For example, farmland can help address climate change through sequestering carbon in soils, especially if the land is farmed well and regenerative practices like cover crops, reduced tillage, rotational grazing, and silvopasture (among others) are used. One California case study reveals that farmland produces much fewer greenhouse gas emissions than land converted to residential or commercial uses does: "Farmland that is converted to other uses emits greenhouse gases at a level 58–70 times greater than if it had remained in farming." Another report in New York shares similar figures, speaking further to the ways that agriculture can be a solution and not a pure stimulant to our climate crisis.[31]

While agricultural land is valuable to our climate and so to all communities—especially if farmed well—it's also useful to each community it sits in. Conserving farmland can provide local-level environmental services, like preserving water quality and protecting wildlife habitat. Because of farmland's vegetated fields and open spaces, it can offer natural flood control, water filtration, and groundwater recharge in ways that impervious, suburban areas cannot. When heavy rains fall on impervious surfaces—like parking lots, paved driveways, and rooftops—water can quickly pool nearby and cause flooding, and pollutants can be carried into local waterways. But when heavy rains fall on fields and wooded areas, the water can more easily be absorbed in place, minimizing flood and pollution concerns. It is true that some farmland—especially on large-scale industrial operations—can leach chemical fertilizers, pesticides, and eroded soils into water bodies. In fact, some research shows that "industrial agriculture is now the largest source of water quality impairments" in the nation. Further, livestock on some farms can harm water quality if they aren't excluded from waterways—or if they are packed into pastures or buildings so tightly that manure deposits are hard to control, as in CAFOs. But many of these water-quality issues, especially on small and midsized farms, can be mitigated by employing proven conservation practices, such as planting riparian buffers. These vegetated areas along swales, creeks, streams, and rivers enable more water absorption and drastically reduce harmful runoff.[32]

Agricultural land serves wildlife too. Bengston and colleagues state that "sprawling development has been implicated as the leading cause of habitat loss and species endangerment in the mainland United States." Elisabeth Johnson and Michael Klemens echo this finding that sprawl leads directly to habitat destruction and fragmentation. By replacing open and wooded fields with houses, roads, and retail stores, development destroys the homes of many creatures and forces them elsewhere, where they stay until development displaces them again. Even when development does not destroy habitat, Patrick Gallagher affirms, it chops up open spaces and disrupts natural migration and foraging patterns.[33] Farmland, on the other hand, offers open spaces and landscapes that benefit many plant and animal species.

Farmland is not perfect for habitat, but it's better than the alternative of sprawling development. It is especially beneficial if the farms have diversified landscapes that incorporate open fields and woods into the same land area—as many nonintensive small and midsized farms do. Even greater benefits can arise if farmers use practices such as adaptive grazing, well-timed mowing and harvests, and sustainable forest management—all of which depend on attunement to and imagination of place—that are considered supportive of wildlife. Turning back to Robertson and Maury Counties, the Tennessee Wildlife Resources Agency agrees that agricultural land can be significant for flora and fauna. In its 2015 State Wildlife Action Plan, it assigned important habitat value to much of these communities' farmland. It did the same in other communities throughout the state. These rankings affirm the significance of farms for nonhuman communities, and they are reflected in other states throughout the region and nation.[34]

In the continued stewardship of small and midsized farms, we see farmers' persistence and perseverance. We see determination, fueled through people-place relationships and stewardship virtues. We also see tangible ways that the continuation and proliferation of these farms and farmers can serve the public good.

For the well-being of farmers, the general population, and the earth itself, we should adopt new agricultural and environmental policies on federal, state, and local levels. We should leverage love for the land into structural change, fostering a farming system in which grounded virtues can thrive. We should make good and virtuous stewardship no longer a sacrifice. Doing so will address the troubles that plague current and future small and midsized farmers. It will protect irreplaceable farmland from development. It will advance racial equity and justice, which have long been denied.

These are all much-needed advancements, and with a thoughtful approach and ground-up guidance, they can be achieved. But it won't be easy. If proposed, the changes described here, along with dozens of others that could lead to progress, will be met with strong resistance, especially by people in power who benefit from maintaining the status quo. That resistance will be well funded and intense. Struggles await anyone who champions the cause of a more just and affectionate agriculture.

In the face of these surefire difficulties and challenges, it's important for policy makers to act in accordance with virtues too. Just as the farmers featured in this book have nurtured imagination, affection, and fidelity when caring for their farms, leaders on federal, state, and local levels will need to cultivate and then legislate with resilient character dispositions that serve the common good. Many virtues could help leaders bring about the changes needed to mend US agriculture and the nation's food system, enhance ecological and environmental sustainability, and support rural communities. Empathy, courage, and hope stand out as especially important.

Rather than resort to lazy stereotypes and assumptions, leaders should actively engage with and listen to small and midsized farmers. They should visit rural communities. But more than just visit, they should spend genuine time there, even if it doesn't at first seem politically advantageous or strategic to do so. These visits must go beyond flying or driving into town for a few days, sporting camouflage caps and just-purchased boots, and taking pictures with the locals who show up. That inauthentic approach serves no one, save the store owner who gets to sell $150 pairs of boots that will only be worn for photo ops.

Leaders should make special efforts to engage with farmers of color and other marginalized communities. While many federal programs like to refer to these groups as "historically underserved," that description is misleading. Lack of support isn't relegated to the past. These groups are *still* underserved, and we need to embrace empathy to listen to their struggles, effectively advocate on their behalf, and ensure they have opportunities and space to advocate for themselves.

This sort of engagement in rural places will take time. It will be hard and, in some instances, uncomfortable. But it will be worth it. Liberal, moderate, and conservative officials alike can, and should, do more to empathize with and serve these smaller-scale farmers and communities. One side of the political aisle tends to pander to the rural US while doing very little to actually support it. That is especially true for diverse rural people and those with low or moderate incomes. The other side tends to ignore rural places, cast quick assumptions and cruel judgments, quickly dismiss real problems, and underinvest in local efforts. Practicing genuine empathy and engagement may begin to change both sides.

With empathy as a guiding force, leaders must then act with courage. Avoiding the polar extremes of cowardice and rashness, they should strategically propose and advance reforms that serve people, places, and the entire public, knowing full and well that these changes may anger influential people on global, national, state, and local levels. They must do all of this with hope—not naïve optimism or wishful thinking but authentic, active, virtuous hope.[35] Recognizing and being honest about very real adversities while continuing to work hard for a better world is paramount. With this sort of empathetic, courageous, and hopeful commitment—an enduring effort that is much like fidelity—we can move toward better days through better ways.

These charges to our representatives are all well and good. But everyone reading this book—and I, the person writing it—must recognize that progress cannot be passive. Change is not entirely up to elected officials, nor is it wholly up to the farmers featured in these pages. It's up to all of us. We've got to work for the world we want to live in.

Some folks are doing that work, and they're doing it well. Whether through individual efforts or participation in and support for organizations, groups, associations, and communities devoted to good stewardship, justice, and vibrant communities, people are pushing for a better future for farming. They are literally putting their money—and, more importantly, their time and energy and ethics—where their mouth is, aiming to engender a better and more equitable food and farming system in places across the nation.

But more of us need to embrace this effort, and we need to do it through a combination of private commitments and public movements. Otherwise, the work will never become personal—as it must. Because we eat, breathe, drink, and live, we are all dependent on good farms and farmers, and we must take responsibility to ensure better stewardship. Urban, suburban, and rural people alike are better off when we amplify, value, and practice imagination of, affection for, and fidelity to place. When we act with the empathy, courage, and hope that we want to see in others. When we cultivate cultures of care.

Heeding lessons from farmers who persist in place, we can embrace these virtues. Rather than give up or get out, we can dig in. Rather than go big, we can go home.

For the sake of people, place, and planet, we should live with love for the land.

Acknowledgments

While writing *Love for the Land*, I often thought about my family's farm and the hard work required of a good farmer. Much of the time, this work can be done alone. But the biggest, most demanding jobs require teamwork.

My family has come together on many occasions to take on tough farm tasks. In these moments, everyone brings their own skills and know-how to the job at hand. Working as one, my family has harvested tobacco and hay crops, fixed a sagging roof on a barn first built in the 1890s, designed and constructed tool sheds, weeded one-acre gardens, sold produce at farmers markets, built fences, rebuilt barn lofts, worked cattle, cleared brush, fixed tractors, planted pumpkins, and so much more. In addition to making jobs easier and final products better, working together makes the job more fun. There is pleasure in doing good work with people you love.

This book is a bit like farm work. It is better because of the many hands, heads, and hearts that touched it. For all the folks who kindly offered assistance and support, edits and encouragement, I am grateful. A few people deserve special recognition.

I want to first thank Hal Manier, whose story anchors the preface. Sadly, Hal died while I was revising this book. He was a friend and inspiration, and he represented our community of Holts Corner well. I miss him and hope this book would make him proud. Thankfully, because of his foresight and fidelity, Hal's land and legacy live on.

Next, I owe a warm thanks to Wendell Berry. For over a decade now, he has inspired me through his writings, words, and actions. He has made me think and reflect, even when we disagree. Beyond offering distant inspiration, I am grateful to him for his written correspondence and for inviting me to his farm for front-porch, back-field, and kitchen conversations. Nurturing a friendship with Mr. Berry has been a great joy.

Berry's writings on stewardship serve as the foundation for this project, but the heart of my work lies in exploring on-the-ground enactments of imagination, affection, and fidelity. I appreciate all the people in rural Tennessee and southern Kentucky who shared their time, stories, and experiences with me. Learning about their love for the land was illuminating. I am especially grateful to people who connected me with a wide range of farmers, community leaders, and farm-service providers. For showing patience and kindness—and for being open and honest in our conversations, even when it was painful—I offer thanks to every participant.

At various points, faculty and staff at Yale University, particularly in the School of the Environment, provided indispensable guidance, support, and feedback. Many thanks to Indy Burke, Michael Dove, Thomas Easley, Brad Gentry, John Grim, Joe Orefice, Maya Polan, Paul Sabin, Oswald Schmitz, Jim Scott, Sara Smiley Smith, Mary Evelyn Tucker, and Libby Wood. I'm also grateful to the Yale Institutional Review Board, who helped ensure my research maintained high ethical standards. Special thanks are due to Amity Doolittle and Justin Farrell. Amity shaped my research methods, offered much-needed written feedback, and helped me better understand environmental justice and injustice. She did all this with compassion and grace, and she believed in me even when I doubted myself. Justin read every word of this work and eagerly discussed many ideas, situations, and opportunities with me. With patience and perseverance, he also helped me navigate conducting qualitative research during a global pandemic, which was no small task. He guided me in the publishing process too. In every way, Justin has been a kind and consistent adviser.

Friends I met at Yale generously shared their support and time. Sadly, I cannot list everyone here, even though scores of people should be recognized. But for their encouragement and enlightening conversations—often around a backyard firepit, in front of an apartment wall covered with bird photos, or on walks through the neighborhood—I am grateful to Lauren Ashbrook, Bridget Barns, Paul Burow, Ryan Clemens, Erin Eck, Cam Humphrey, Colin Korst, Katie McConnell, Jenna Musco, Bill Pedersen, Ashley Stewart, and Anelise Zimmer. Because they also read parts of the book and offered feedback and advice, extra thanks go to Brian Basso and Jesse Williams.

Yale University Press believed in this book and helped bring it to life. While I am grateful to the entire staff at the press, I especially appreciate the thoughtful suggestions, patience, and guidance of Jeff Schier, Elizabeth Sylvia, and Jean Thomson Black. I'm grateful, too, for the reviewers who offered helpful comments—especially Norman Wirzba and Brian Donahue. Their thoughts and recommendations made the book better.

For providing information about Century Farms in Tennessee, I thank Antoinette van Zelm at Middle Tennessee State University's Center for Historic Preservation and Donna Baker at MTSU's Albert Gore Research Center. I also owe special thanks to staff at The Land Trust for Tennessee—particularly my friends and former colleagues Emily Parish and Mike Szymkowicz—for connecting me with farmers in my field sites. My thanks go to the Harry S. Truman Scholarship Foundation for their generous, steadfast support as well. Others offered support too, whether in the form of reading parts of the manuscript, engaging in deep discussions, or giving me space to write and revise. I'm grateful to Gene and Julie Adolph, Amanda Cather,

Rachel Cheek, Laura Freeman, Erica Goodman, Matthew Jakes, Jamie Mierau, Lee Miller, and John Piotti, among others.

Finally, I want to thank my family. My parents, Ken and Angela, supported me throughout this project, offering both encouragement from afar and hands-on help when I was in Tennessee conducting field work. Most importantly, they have shown me the power of love and truth, service and sacrifice, perseverance and loyalty through a lifetime of examples. My brothers, Michael and Patrick, gave essential support too. On several occasions while I was home, Patrick and I worked together on farm-focused projects. The sagging barn roof mentioned earlier, for example, would not have been fixed without his efforts and skills. Shared tasks on the farm were a source of joy, purpose, and inspiration. Michael offered feedback on every chapter and encouraged me during tough times. He shared expertise on hope and character. And as one of the hardest working people I know, he inspired through example. He was, and is, always eager to help. For that and more, I'm grateful.

My last and most important note of thanks goes to my best friend and wife, Regan. She has read this entire work at least four times and combed it for copyedits twice. She has spent dozens, if not hundreds, of hours in conversation with me about stewardship virtues, farmland loss, justice, equity, agriculture, and ethnographic research. She has taught me about empathy and affection through her steady example. Even when we were twelve hundred miles apart, Regan was right there with me. For her grace, her trust, her fidelity, and her love—and for helping me and so many others flourish—I give my deepest gratitude.

Notes

1. The Virtues of Imagination, Affection, and Fidelity

1. National Endowment for the Humanities, "2012 Jefferson Lecture with Wendell Berry."

2. Berry may disapprove of the descriptor "environmental." In a 2019 interview with Amanda Petrusich of the *New Yorker*, Berry said, "The thing that worries me very much is how much language we're using now that is so abstract as to require no thought at all. . . . People speak of 'the environment.' They don't know what they're talking about. 'The environment' refers to no place in particular. We're alive only in some place in particular" ("Going Home with Wendell Berry"). Berry prefers the descriptor "ecological" over "environmental" because of the former's necessary locality, but here, I find that "environmental" is helpful for avoiding confusion and maintaining clarity for a broader audience.

3. Berry, "It All Turns on Affection," 14–15, 32. Other agrarian and environmental authors and leaders, including Henry David Thoreau, George Washington Carver, Liberty Hyde Bailey, Aldo Leopold, Rachel Carson, Annie Dillard, Drew Lanham, Wes Jackson, and bell hooks have, to varying degrees, also touched on the importance of intimacy with the earth. Further, in a 1968 meeting of the International Union for the Conservation of Nature and Natural Resources, the Senegalese forester Baba Dioum stated, "In the end we will conserve only what we love; we will love only what we understand; and we will understand only what we are taught." Others, including Indigenous leaders, have also spoken of the importance of close connections with place, and still others—like Harriet Tubman, whose ecological knowledge of swamps, woods, fields, and more helped her lead enslaved peoples to freedom (Taylor, *Rise of the American Conservation Movement*)—practiced eco-attunement. While a historical and contemporary tracing of place-based attachment, attention, and affection would be interesting and insightful, this chapter is primarily concerned with Wendell Berry's stewardship virtues. Berry's focus on personal stewardship does not mean that he is not an advocate for policy solutions. In fact, he has advocated for policy changes that

would better serve people, agriculture, and the earth. Along with Wes Jackson, he has championed a "fifty-year farm bill"—rather than the typical five-year bill—that places stewardship and equity at the heart of agricultural legislation (Berry and Jackson, "50-Year Farm Bill").

4. Johnson, "We Must Reject"; Berry, "It All Turns on Affection," 10; Dunlap, "Conventional versus Alternative Agriculture"; Goldschmidt, *As You Sow*; Hribar, *Understanding Concentrated Animal Feeding Operations*; Kimbrell, *Fatal Harvest Reader*; Kremen, Iles, and Bacon, "Diversified Farming Systems"; Lobao, *Locality and Inequality*; Lyson, Torres, and Welsh, "Scale of Agricultural Production"; Lehner and Rosenberg, *Farming for Our Future*, 24–25, 40–59, 164; Sherrick, "Prose to Policy"; Spangler, Burchfield, and Schumacher, "Past and Current Dynamics"; Stoll, *Larding the Lean Earth*; Olmstead, *Uprooted*; C. Thompson, *Going over Home*. In the essay "Think Little," Berry recognizes that many farmers who act as exploiters do so because larger economic and agricultural forces push them to do so (85–86).

5. Berry, "It All Turns on Affection," 10, 11, 32–33; Leopold, *Sand County Almanac*; K. Smith, *Wendell Berry and the Agrarian Tradition*, 56.

6. Annas, *Intelligent Virtue*, 8. Philosophers describe virtue in various ways. For this book, I have constructed a definition that I think is accurate and appropriate.

7. Zagzebski, *Virtues of the Mind*, 84, 88–89, 89–102.

8. Annas, *Intelligent Virtue*, 8–10.

9. Zagzebski, *Virtues of the Mind*, 120; Aristotle, *Nicomachean Ethics*, 26. Some scholars and theologians argue that virtues can be infused by God's grace and are thus not entirely acquired by individuals. I approach this element of virtue in an Aristotelian light and choose to amplify the importance of habituating oneself toward virtues and ethical actions.

10. Annas, *Intelligent Virtue*, 13–15.

11. Roberts, "Will Power and the Virtues," 235; Musonius Rufus, "Lecture No. 11," 189–190.

12. Zagzebski, *Virtues of the Mind*, 85.

13. Cafaro, "Thoreau, Leopold, and Carson," 3. Rights-based and legislative approaches to environmentalism have proven successful on large scales, and I am not critiquing their use. In fact, I applaud them. However, the appeal of virtue-based environmentalism is that it helps encourage individual environmental commitments. Personal environmental efforts are of the utmost importance in confronting the ecological challenges we have created. Berry argues for this intimate environmentalism in great detail in his essay "Think Little." Sandler, *Character and Environment*, 5; van Wensveen, *Dirty Virtues*; P. Thompson, *Spirit of the Soil*; Bai, Chang, and Scott, *Book*

of Ecological Virtues; Kawall, *Virtues of Sustainability*. In "Respect for Nature," Christine Cuomo offers a thorough discussion of Indigenous environmental virtues.

14. K. Smith, *Wendell Berry and the Agrarian Tradition*, 214; van Wensveen, *Dirty Virtues*, 12; P. Thompson, *Spirit of the Soil*, 100. Berry avoids academic jargon in his discussion of virtues, I believe, because he feels that these discussions can get stuck in the ivory towers and serve no practical purpose. Virtues should not exist only in the mind—they must be acted on in reality. In "The Body and the Earth," Berry writes that "a purposeless virtue is a contradiction in terms" (125).

15. M. Lamb, "Difficult Hope."

16. Berry, "It All Turns on Affection," 14.

17. Berry, "Imagination in Place," 2, 10.

18. Berry, "It All Turns on Affection," 14; Sutterfield, *Wendell Berry and the Given Life*, 123.

19. Bilbro, *Virtues of Renewal*, 25–42; Dodson Gray, "Critique of Dominion Theology," 80. A 2018 documentary on Berry's life and writing and the ways that farming has changed in his native Kentucky is titled *Look and See: A Portrait of Wendell Berry*. In it, Berry discusses the differences between "looking" and "seeing." More can be learned about the film by visiting https://lookandseefilm.com.

20. Berry, "Renewing Husbandry," 94–97; Berry, *Unsettling of America*, 9. In many academic circles, this type of scientific or quantitative knowledge is described as "positivist" and implies that there is one correct way of securing real, usable, legitimate knowledge. Berry, and I, would challenge this approach. An example of the destructive generality of scientific knowledge can be found in examining the Green Revolution. Scientists and development practitioners imposed Western knowledge on local peoples in the Global South, replacing local imagination with technology. This same science was applied in the United States, where more traditional approaches to farming—such as the integration of livestock and crops on the same fields to improve soil health and fertility, for example—were replaced by chemical-intensive processes. While the Green Revolution did increase yields of many crops, it also resulted in substantial harm to many communities, and it obliterated generations of deep cultural knowledge about agriculture.

21. Scott, *Seeing Like a State*, 313, 318, 324, 333–341. Berry also discusses farming as an art form in many of his essays, poems, and novels. This depiction is especially clear in his book *The Art of Loading Brush*.

22. Berry, *Unsettling of America*, 35. Here, a "better farmer" is understood as one who is focused on being an excellent, loving caretaker for the land and community while still providing for themselves and their family. In a 2019 interview with Amanda

Petrusich in the *New Yorker*—cited earlier—Berry himself said that a "good farmer" is "one who brings competent knowledge, work wisdom, and a locally adapted agrarian culture to a particular farm that has been lovingly studied and learned over a number of years" ("Going Home with Wendell Berry").

23. Wiebe, *Place of Imagination*, 16.

24. Wirzba, "Economy of Gratitude," 152; Berry, "It All Turns on Affection," 14–15; hooks, *All About Love*.

25. DiEnno and Thompson, "For the Love of the Land," 67–72.

26. Verduyn and Lavrijsen, "Which Emotions Last Longest and Why," 119.

27. Berry, "It All Turns on Affection," 33; Wiebe, *Place of Imagination*, 108.

28. Berry, *Gift of Good Land*, 180–182.

29. Berry, "Conservationist and Agrarian," 74–75.

30. Wirzba, "Economy of Gratitude," 152; Berry, *Gift of Good Land*, 155–160; Berry, "Defense of the Family Farm," 33–34. For an example of Berry's discussion of a new farmer beginning on new land and still cultivating imagination and affection, see Oehlschlaeger, *Achievement of Wendell Berry*, 19–20.

31. Berry, "It All Turns on Affection," 14, 33.

32. Berry, *Unsettling of America*, 125, 127.

33. Wiebe, *Place of Imagination*, 83; Berry, "People, Land, and Community," 186; K. Smith, *Wendell Berry and the Agrarian Tradition*, 144; Berry, *Unsettling of America*, 128.

34. Berry, *Unsettling of America*, 125; Berry, "It All Turns on Affection," 35; Berry, "Think Little," 86.

35. Berry, "Imagination in Place," 1; Berry, "Native Hill," 5; Berry, "It All Turns on Affection," 14.

36. Berry, "Native Hill," 5.

37. Berry, 5–6, 7.

38. Berry, "It All Turns on Affection," 13.

39. Berry, "Health Is Membership," 106.

40. Berry, *Unsettling of America*, 9.

41. Berry, *Gift of Good Land*, 206.

42. Berry, 208.

43. Berry, 209.

44. Planting fence row to fence row was another of Secretary Earl Butz's commands to farmers in the 1970s. Berry, 227–228.

45. Berry, 227–235.

46. Berry, 232 (emphasis added).

47. Berry, "Defense of the Family Farm," 47.

48. Berry, *Gift of Good Land*, 250, 253, 259–263.

49. Berry, 261–263.

50. Dudley, *Debt and Dispossession*; Salamon, *Newcomers to Old Towns*; Olson and Lyson, *Under the Blade*; Moroney and Castellano, "Farmland Loss and Concern in the Treasure Valley"; Bell, *Farming for Us All*; LeVasseur, *Religious Agrarianism and the Return of Place*; Larmer, "Cultivating the Edge."

51. In "Ethnography and Ethics," Jung Lee notes the role that formal ethnography can play in observing and communicating practical virtues and argues that this empirical analysis should be used more often. Lee states that ethnography can offer much-needed insights into lived ethics and virtues.

52. Berry, "Think Little," 87. In "Exemplarist Environmental Ethics," Alda Balthrop-Lewis offers an excellent articulation of this idea: "Living well is pushing as *you* can from where *you* are for the future of the world we share, it is loving some place and some people well enough—my mother taught me and my cousins and my nieces and nephews to love the sea grass beds and the communities that rely upon them—that you can mourn well the harms they undergo, work like hell to protect them, delight in the goods that remain, and seek systemic political reform for the good of all" (546).

53. Sherval et al., "Farmers as Modern-Day Stewards."

54. Jager, *Fate of Family Farming*, 195; Berry, "It All Turns on Affection," 33. In "The State of US Farm Operator Livelihoods," Burchfield et al. discuss the costs of our modern industrial agricultural system: "The focus on producing calories and consumer goods as cheaply as possible . . . has meant that the true costs of food production have been externalized—whether through the reduced nutritional content of food, environmental degradation, unethical labor practices, or . . . the livelihoods of those who operate US farms" (1).

55. Berry, "Farmland without Farmers"; Donahue, *Go Farm, Young People*; Jackson, "Prologue," 4–5. For more on defining scale, see note 2 in chapter 2.

56. B. Lamb, *Overton Park*; Berry, "Think Little," 88–89; Bailey, *Holy Earth*, 36. Here, Liberty Hyde Bailey writes, "If it were possible for every person to own a tree and to care for it, the good results would be beyond estimation."

2. "The Hard Thing Is Keeping the Land"

1. Baker, *Bulldozer Revolutions*; Coulthard, "Changing Landscape of America's Farmland"; Rome, *Bulldozer in the Countryside*; Thompson and Prokopy, "Tracking Urban Sprawl"; Freedgood et al., *Farms under Threat*; Hunter et al., *Farms under*

Threat 2040. Although Tennessee was ranked fourth in the nation for farmland loss from 2001 to 2016, the state jumps to third when projecting future farmland loss over the next two decades, just behind Texas and North Carolina. Estimates by Hunter et al. show that under current development trends, Tennessee could sacrifice over a million additional agricultural acres—or more than 8 percent of its total farmland—by 2040 (22).

2. Establishing parameters for farm size is a difficult task, and people approach it in different ways. Some researchers and officials, for example, use total farm income or sales to assign a size category to farms. But in *Family Farming: A New Economic Vision*, Marty Strange argues against using this metric to describe farm size, largely because of price discrepancies between years, differences in profit margins between farm type, variations within farm size group, and bunching toward the low end of each classification's income spectrum (69–71). To be sure, using acreage as the primary metric also poses problems. What is considered a small farm in one community is massive in another. Variation between farm type and primary product is also a concern when using acreage as the guiding metric. For example, a 120-acre tobacco farm might be considered large, while a 120-acre dairy farm is seen as small. A 1,000-acre sugarcane farm may be one of the smallest around, while a 1,000-acre vegetable farm would be massive. Though subjective and imperfect, the range of 50–499 acres tries to capture entities that are large enough to be considered farms (i.e., more than a "hobby farm") yet small enough to still be considered midsized instead of large, and it is attuned to the local agricultural dynamics of Middle Tennessee. This specific range may not accurately define small and midsized farms in other counties, states, regions, or nations.

3. Roberts, "Will Power and the Virtues." Although their analysis is anchored in Florida, Joy Goodwin and Jessica Gouldthorpe offer a helpful overview of the challenges that smaller-scale farmers face across geographies in "Small Farmers, Big Challenges." I explore some of these struggles briefly later in this chapter.

4. On the widespread acknowledgment of the importance of the US Census of Agriculture's data, Spangler, Burchfield, and Schumacher write, "The Census is the only source of detailed county-level agricultural data that is collected, tabulated, and published using a uniform set of definitions and methodology. Thus, the Census is considered the most complete count and measurement of U.S. farms, operators, and ranches in the U.S. . . . It provides the most comprehensive, open-source record of historical U.S. agricultural data" ("Past and Current Dynamics," 3).

5. Bengston et al., "Analysis of the Public Discourse," 745. In *From Sprawl to Sustainability: Smart Growth, New Urbanism, Green Development, and Renewable*

Energy, Robert Freilich, Robert Sitkowski, and Seth Mennillo offer an explanation for the prevalence and destructive power of sprawling development throughout the United States: "Sprawl is accelerated by the predominant American desire for an imagined rural lifestyle coupled with an urban income, while ignoring the catastrophic economic, environmental, traffic, and fiscal impacts that such living patterns create. . . . [People are] imitating rural life, while simultaneously demanding a full range of urban public services, from daily police patrols, sewer, and public water to fire hydrants, swift emergency response, and school bus stops in front of the home" (23). See also Olson and Lyson, *Under the Blade*.

6. Ewing and Hamidi, *Measuring Sprawl 2014*. It is worth noting that Tennessee's six major metropolitan areas—Chattanooga, Clarksville, Kingsport-Bristol, Knoxville, Memphis, and Nashville—all finished in the bottom 12 percent in the sprawl index, meaning that Tennessee's major cities as a whole are among the most sprawling in the nation. Jeong, "Nashville Approves New Budget."

7. Szlanfucht, "How to Save America's Depleting Supply," 334; C. Francis et al., "Farmland Conversion"; J. White, "Beating Plowshares into Townhomes," 116.

8. Brown and Schafft, *Rural People and Communities*, 181; Tolbert et al., "Civic Community in Small-Town America." Jane Preston explores this notion further in "Community Involvement in School," a 2013 article on the lack of social cohesion in bedroom communities. Preston specifically anchors her research in education studies and shows how and why both parents and children tend to be less involved with schools and supporting entities in bedroom communities. Preston quotes one research participant as saying, "For so many people, the city is their focus. The city is where they bring their kids for a lot of events. The city is where they work. The city is where they do their shopping and socializing. So I guess, in a sense, we don't need each other as much as communities which are further away from the city" (426). Many others shared the same sentiments.

9. Salamon, "From Hometown to Nontown," 1; Salamon, *Newcomers to Old Towns*; Pilgeram, *Pushed Out*; Sherman, *Dividing Paradise*.

10. Irwin, Cho, and Bockstael, "Measuring the Amount and Pattern of Land Development"; Freedgood et al., *Farms under Threat*, 28; American Farmland Trust, "Tennessee."

11. Hanson, Hendrickson, and Archer, "Challenges for Maintaining Sustainable Agricultural Systems," 328–332; Bunge, "Supersized Family Farms"; Semuels, "They're Trying to Wipe Us Off the Map"; M. Young, "Keep It in the Family"; Whitt, Todd, and MacDonald, "America's Diverse Family Farms," 7.

12. Sumner, "American Farms Keep Growing," 149.

13. MacDonald, Hoppe, and Newton, "Three Decades of Consolidation," iii; Ferdman, "Decline of the Small American Family Farm."

14. MacDonald, Hoppe, and Newton, "Three Decades of Consolidation," iii; Koerth, "Big Farms Are Getting Bigger."

15. USDA, "Farm Household Well-Being." The Federal Reserve's Inflation Calculator shows that $1 in 1975 is equal to $5.47 in 2022, demonstrating the ways in which the USDA's farm definition has not kept pace with inflation.

16. Wozniacka, "Is It a Farm if It Doesn't Sell Food?"

17. Koerth, "Big Farms Are Getting Bigger."

18. Here, I include "impassioned" as a reference to Berry's amateur ethnographies. Work done with love can, I think, lead to important results.

19. Data from the 2017 Census of Agriculture show that over 20 percent of farms in Robertson County do not have internet access.

3. Farmland Loss and Big Agriculture

1. "Carl" is a pseudonym. Based on the recommendation of Yale University's Institutional Review Board, I assigned pseudonyms to all interview participants. Each false name was selected at random from the Social Security Administration's list of the top one hundred names for males and females over the past one hundred years.

2. N. Young, "Confusion over Greenbrier-Area Development"; Freedgood et al., *Farms under Threat*, 4, 21, 28.

3. American Farmland Trust, "Tennessee."

4. These descriptions of "bedroom towns" align with the observations and analyses made by Sonya Salamon in *Newcomers to Old Towns: Suburbanization of the Heartland*. They also align with research in Tom Daniels's *When City and Country Collide: Managing Growth in the Metropolitan Fringe* and Ryanne Pilgeram's *Pushed Out: Contested Development and Rural Gentrification in the U.S. West*.

5. Generational turnover is an issue across the nation. Because of the increasing average age of farmers, Julia Freedgood et al. of the American Farmland Trust estimate in *Farms under Threat* that about 370 million acres of agricultural land will change hands in the next two decades. The organization recommends making generational turnover a focal point in agricultural policy.

6. For more information on and stories about the devastating suicide trend, see Semuels, "They're Trying to Wipe Us Off the Map"; *Successful Farming*, "Farmer Suicides Today vs. 1980s Farms Crisis"; Wedell, Sherman, and Chadde, "Midwest Farmers Face a Crisis"; Weingarten, "Why Are America's Farmers Killing Them-

selves?" For more on American farming's unprecedented current levels of farm debt, see Burchfield et al., "State of US Farm Operator Livelihoods," 15–16.

7. Some scholars do recognize the importance of qualitative research and emotion in the realm of sprawl, farmland loss, and changing rural communities. For several strong and insightful examples, see Sonya Salamon's *Newcomers to Old Towns*, Kathryn Dudley's *Debt and Dispossession*, Melissa Walker's *Southern Farmers and Their Stories*, Jennifer Sherman's *Dividing Paradise*, Ryanne Pilgeram's *Pushed Out*, and Justin Farrell's *Battle for Yellowstone*.

8. In *Family Farming: A New Economic Vision*, Marty Strange briefly touches on how farmland loss stirs difficult emotions for people in rural communities, especially other farmers. "Television can let you see farmers going broke," he says, speaking of folks who observe this phenomenon from afar but do not live in rural places. "But it can't make you feel the pain of losing a neighbor or watching a friend's life collapse" (15).

9. Brian's family was not alone in being affected by the Farm Crisis. Several participants described losing or almost losing some or all of their land. For a thorough exploration of the Farm Crisis, see Kathryn Dudley's *Debt and Dispossession* and Barry Barnett's "The U.S. Farm Financial Crisis of the 1980s."

10. Semuels, "They're Trying to Wipe Us Off the Map."

11. For more on the culture of growing tobacco, see Wynne Wright's "Fields of Cultural Contradictions: Lessons from the Tobacco Patch." While the article is anchored in Kentucky and focuses on burley tobacco production, the discussion of tobacco's cultural importance for some farmers is comparable to the dynamics in Robertson County, Tennessee.

12. Swanson, *Golden Weed*, 251.

13. Balvanz et al., "Next Generation," 76–77.

14. The farmer's portraying of "welfare" here in a negative light is unfair to low-income people who rely on this assistance. However, data does show that big farmers do get the vast majority of government financial support. For example, a recent article by Donald Carr and Anne Schechinger shows that the overwhelming majority of COVID-19 pandemic relief money in the agricultural field went to the nation's largest, wealthiest farmers. By extension, that also means that these payments went almost exclusively to white farmers. Further, one interviewee even noted that some larger farmers he knows employ someone to explore subsidy programs and maximize government payments. For more on this issue, see Strange, *Family Farming*; MacDonald, Hoppe, and Banker, "Growing Farm Size"; and McFadden and Hoppe's "Evolving Distribution of Payments."

4. Neither Getting Big nor Getting Out

1. Louke van Wensveen writes in her foundational book *Dirty Virtues: The Emergence of Ecological Virtue Ethics*, "We need not always rely on individual virtues to achieve balance. We can rely on a network of virtue relations. This view of the life of virtue nicely corresponds with the general emphasis in eco-literature on relational or holistic modes of thinking, acting, and being" (15). The interdependence of imagination, affection, and fidelity aligns with her vision of the ecological virtues.

2. Rebanks, *Pastoral Song*, 26.

3. In an interview with Amanda Petrusich of the *New Yorker*, Berry tells a similar story about a small farmer who knew how to get the most out of his corn crop in a poor economic market. Rather than harvesting all of the corn and sending it to market, the farmer turned his pigs into the field and let them eat their fill. The farmer's pigs fared particularly well that year, earning a decent income (Petrusich, "Going Home with Wendell Berry"). Both James's and Berry's stories reveal economic advantages of finely tuned imagination.

4. Speaking of his family's farm in "A Native Hill," Berry writes, "I had come to be aware of it as one is aware of one's body; it was present to me whether I thought of it or not" (5).

5. Robert and his family still raise a small amount of tobacco—roughly fifteen acres. They focus on quality over quantity and sell some of their leaves for high-end purposes, like fine cigar wrappers.

6. Berry, "Notes from an Absence," 42.

7. To see the rest of the FFA Creed, see FFA, "FFA Creed."

8. Berry, "Conservationist and Agrarian," 74–75.

9. Burchfield et al., "State of US Farm Operator Livelihoods," 18.

10. Petrusich, "Going Home with Wendell Berry."

11. Because landowners are voluntarily giving up value in order to conserve their land forever—which provides a public good, especially in the form of environmental benefits—Congress has enabled people who donate conservation easements to qualifying land trusts to take a limited tax deduction. For more, see Land Trust Alliance, "Frequently Asked Questions." Melissa described working in the past to protect land from development with some people who viewed the act of doing so as a pure business decision. These are the "exceptions" she mentioned.

12. For an interesting agrarian interpretation of Christianity and the Bible, see Davis, *Scripture, Culture, and Agriculture* and LeVasseur, *Religious Agrarianism and the Return of Place.*

13. Leopold, *Sand County Almanac*, xix.

14. James's outlook—particularly the element that focuses on long-term interdependence between people and the earth—also aligns with many Indigenous traditions.

15. Berry, "Starting from Loss," 88.

5. Systemic Struggles

1. "Walter" is a pseudonym. As with the names of Robertson County participants, the names of participants from Maury County have been replaced by pseudonyms from the Social Security Administration's list of the top one hundred names for males and females over the past one hundred years.

2. It should be noted that, for people of all races, the US Census of Agriculture counted producers differently in 2017 than in years past. Recognizing that there is often more than one person making farming decisions for each farm—such as a married couple or multiple siblings—the census allowed people to list multiple farmers for a farm. This means that although there are sixty-one Black farmers listed for Maury County, there may not be sixty-one Black-owned farms.

3. Daniel, *Dispossession*; Horst and Marlon, "Racial, Ethnic and Gender Inequities"; Reid and Bennett, *Beyond Forty Acres and a Mule*; Humphrey, "Centering a Population Overlooked in Agriculture"; Touzeau, "Being Stewards of Land Is Our Legacy."

4. As with farms in Robertson County, I am classifying "small and midsized farms" in Maury County as those between 50 and 499 acres in size. Again, this is an imperfect measurement, but it seems appropriate here.

5. When interviewing these participants—all of whom willingly agreed to speak with me—I followed the same procedures that I used in Robertson County. Each participant was provided a copy of the verbal consent form. Interviews were semi-structured and followed a prepared protocol. I tried to act with empathy and understanding before, during, and after each conversation. And interview data was carefully coded using NVivo software. To ensure safety during a period when COVID-19 cases were rising in Tennessee, all interviews were conducted over the phone. In reference to precedents for small sample sizes, Peter Balvanz et al. engaged five participants in their community-based participatory research with Black farmers in North Carolina ("Next Generation"), while Leslie Touzeau interviewed seven participants in her work on an emerging population of young Black farmers ("Being Stewards of Land Is Our Legacy").

6. Spring Hill is located on the border of Maury and Williamson Counties, so a portion of the town's growth does not technically apply to Maury County. Williamson County has also grown rapidly and converted much farmland.

7. Land area covered by farms that were between 1 and 49 acres in size decreased from 16,145 acres to 15,907 acres, a drop of 238 acres.

8. Hinson and Robinson, "We Didn't Get Nothing," 283–285; Taylor, "Black Farmers in the USA and Michigan," 51.

9. Horst and Marlon, "Racial, Ethnic and Gender Inequities," 3; Carney, *Black Rice*; Hinson and Robinson, "We Didn't Get Nothing," 285; Littlefield, *Rice and Slaves*.

10. Tolnay, *Bottom Rung*, 5. Megan Horst and Amy Marlon write that these four million enslaved people accounted for roughly 16 percent of the nation's wealth in 1860. In today's numbers, that would equate to roughly $10 trillion. In *Black Faces, White Spaces: Reimagining the Relationship of African Americans to the Great Outdoors*, Carolyn Finney, citing research from Elizabeth Blum's "Power, Danger, and Control," shows that careful stewardship of the land—both agricultural areas and "wilder" places, like woods and waterways—was important for enslaved Africans: "Africans believed in 'good use' of the land and the connection between the health of the land and their community" (58).

11. Glymph, *Women's Fight*; Huebner, *Liberty and Union*; J. Smith, *Black Soldiers in Blue*; Horst and Marlon, "Racial, Ethnic and Gender Inequities," 3; Reynolds, *Black Farmers in the Pursuit of Independent Farming*; Tolnay, *Bottom Rung*, 9.

12. As quoted in Darity, "Forty Acres and a Mule in the 21st Century," 660; Pennick, Gray, and Thomas, "Preserving African American Rural Property,"154; Hinson and Robinson, "We Didn't Get Nothing," 286.

13. Taylor, "Black Farmers in the USA and Michigan," 51–52. In reference to the "limited number" of African American people who were offered land, Valerie Grim writes in "Between Forty Acres and a Class Action Lawsuit" that "the majority of newly freed families did not receive either a mule or forty acres." Moreover, Grim explains that few Black people "gained admittance to state-based land grant colleges established by the Morrill Act of 1862. Not until 1890 did the USDA respond to African American petitioners who sought access to public education and experiment stations and the right to participate in other USDA programs with passage of the Morrill Act of 1890" (271). While some policies had good intentions, Civil War agricultural legislation did little to actually serve Black farmers.

14. Darity, "Forty Acres and a Mule in the 21st Century," 660–661; Hinson and Robinson, "We Didn't Get Nothing," 286; Schultz, "Benjamin Hubert," 92.

15. Castro and Willingham, "Progressive Governance"; Miller, "Land and Racial Wealth Inequality"; Du Bois quoted in Hinson and Robinson, "We Didn't Get Nothing," 286.

16. Equal Justice Initiative, "Community Remembrance Project."

17. Agee and Evans, *Let Us Now Praise Famous Men*; Rosengarten, *All God's Dangers*; Hinson and Robinson, "We Didn't Get Nothing," 286–287; Tolnay, *Bottom Rung*, 119–123. Wendell Berry writes that Nate Shaw was a remarkable person with great character, a person whose mind and body were "unified" and who acted on virtuous principles. Calling Shaw a "superior" man, Berry also notes that "his loyalty to his place made him a conservationist," drawing clear parallels to fidelity ("Remarkable Man," 22–28).

18. Taylor, "Black Farmers in the USA and Michigan," 52; Schultz, *Rural Face of White Supremacy*, 46; Petty, "Jim Crow Section of Agricultural History," 26; Canaday and Reback, "Race, Literacy, and Real Estate Transactions"; Ferrell, "George Washington Carver"; Hinson and Robinson, "We Didn't Get Nothing," 288; Jones, "Thomas M. Campbell"; Kremer, *George Washington Carver*; Ruffin, *Black on Earth*, 77–85; Schechter, "On Violence in the South"; Schultz, "Benjamin Hubert"; M. White, *Freedom Farmers*; Daniel, "African American Farmers and Civil Rights," 3; Gilbert, Sharp, and Felin, "Loss and Persistence," 2.

19. Minor, "Justifiable Pride"; Schultz, "Benjamin Hubert"; Womack, "Black Power"; Taylor, "Black Farmers in the USA and Michigan," 55–57; Conklin, *Revolution Down on the Farm*; Hinson and Robinson, "We Didn't Get Nothing," 290; Wood and Gilbert, "Returning African American Farmers to the Land," 44.

20. Tolnay, *Bottom Rung*, 20–24, 120–167; Pennick, Gray, and Thomas, "Preserving African American Rural Property," 155; Gilbert, Sharp, and Felin, "Loss and Persistence," 10.

21. Daniel, *Dispossession*.

22. Taylor, "Black Farmers in the USA and Michigan," 61.

23. Grim, "Between Forty Acres and a Class Action Lawsuit"; Taylor, "Black Farmers in the USA and Michigan," 58–62. In order to be eligible for funds from *Pigford*, Black farmers had to prove that they were the direct victims of racial discrimination. For many folks, being able to provide clear and indisputable proof was a major difficulty and has made receiving allocated funds a challenge. In *Just Harvest: The Story of How Black Farmers Won the Largest Civil Rights Case against the U.S. Government* (2021), Greg Francis, who served as one of the lead counsels in this case, offers an in-depth and behind-the-scenes exploration of the *Pigford* cases.

24. Sewell, "There Were Nearly a Million Black Farmers"; USDA, "2017 Census of Agriculture Highlights."

25. Dyer and Bailey, "Place to Call Home," 317.

26. Albritton and Williams, "Disasters Do Discriminate," 432–433; Castro and Willingham, "Progressive Governance"; B. Lamb, "Understanding Heirs' Property";

Presser, "Their Family Bought Land"; M. Young, "Keep It in the Family"; Bailey and Thomson, "Heirs Property"; D. Francis et al., "Black Land Loss."

27. Albritton and Williams, "Disasters Do Discriminate"; Dyer and Bailey, "Place to Call Home," 318; USDA, "Heirs' Property Landowners."

28. Gilbert, Sharp, and Felin, "Loss and Persistence," 8; B. Lamb, "Understanding Heirs' Property."

29. Presser, "Their Family Bought Land."

30. Penniman, *Farming While Black*; M. White, *Freedom Farmers*. For more about Africulture, visit the home page of Carter Farms: https://thecarterfarms.com.

31. NPR, "Theft at a Scale That Is Unprecedented"; Weissman, "Debt Long Overdue"; Adams and Tucker, "HBCUs Cheated out of Billions."

32. Castro and Willingham, "Progressive Governance."

33. Charles, "Debt, Racism, and Fear of Displacement."

34. In a frequently cited study, Derald Sue et al. describe racial microaggressions as "brief and commonplace daily verbal, behavioral, or environmental indignities, whether intentional or unintentional, that communicate hostile, derogatory, or negative racial slights and insults toward people of color. Perpetrators of microaggressions are often unaware that they engage in such communications when they interact with racial/ethnic minorities" (271).

35. Parvini, "Nashville's Southern Hospitality."

36. Tennessee's lack of state income tax is offset by its sky-high sales tax, which is the highest in the nation. High sales taxes often put a disproportionate burden on poorer people.

37. It is worth briefly exploring Walter's reference to "ticky-tacky houses." Pete Seeger's 1963 hit song "Little Boxes," written and originally performed by Malvina Reynolds in 1962, critiques suburban sprawl, the destruction of close-knit communities, and conformity in the US. It describes the materials that many suburban homes are built out of as "ticky-tacky," a weak material that will crumble over time. The entire song provides a fascinating social commentary.

38. Moran, "Beginning Farmers"; Bowlin, "Joke's on Them"; Fairbairn, *Fields of Gold*.

39. In *Belonging: A Culture of Place*, bell hooks distinguishes between the harmful health effects of tobacco and the value of the plant itself. As someone who grew up raising tobacco in rural Kentucky, hooks describes the beauty and familial connections that defined the growing and harvesting process. In tobacco, she knew cultural and economic value (106–115). Barbara Kingsolver also writes about the tensions between the importance of tobacco for smaller-scale family farms and the adverse health

impacts stemming from this cancer-causing crop (foreword to *Essential Agrarian Reader*, ix–xii).

40. Samuel has frequently participated in the annual mule pull at Columbia's Mule Day celebration with his animals. Mule pulling is a sport in which competitors see whose mules can pull the most weight on a sled the furthest on a dirt track. While Samuel is good at this sport, he said his brother—who is a many-time mule-pulling champion—is even better.

6. "I Wouldn't Take Nothing for It"

1. Coincidentally, the personal experience I share here mirrors one told by Berry in his 1978 essay "Looking Ahead," found in *The Gift of Good Land*.

2. In a comparable story that the renowned ecologist and biologist Drew Lanham relates in his memoir, *The Home Place: Memoirs of a Colored Man's Love Affair with Nature*, he cites his love for his family's farm and mentions the affection that he and others felt for the cattle on their land. Like Ronald, Lanham notes the contentment he felt in knowing that the cows were cared for (117–123).

3. The word Samuel uses to describe his family's land—"homeplace"—is the same term used by the aforementioned Drew Lanham when talking about his family's farm in South Carolina. Lanham, who also comes from a Black farming family, credits this farm with much of his passion for the outdoors and describes it intimately in his memoir (*Home Place*).

4. In "The Terror of Land Loss, the Dream of Finding Home," Martin E. Marty offers powerful discussions of the history and spiritual impact of "home" for rural Black Americans and others.

5. Paralleling Ruth's comment—as well as an earlier story from Walter about his grandmother's death and the family farm work that followed—Berry offers a powerful message on the importance of active responsibility for place in his short story "Stand by Me." After an unexpected, heartbreaking death in the family, the story's narrator notes that despite the overwhelming pain and loss, the farm still needs care: "This place is not a keepsake just to look at and remember. You can't stop just because you're carrying a load of grief and would like to stop. . . . This world was still asking things of us that we had to give" (111–112).

6. The tradition and power of Black land stewardship and belonging is well represented by the commitment of the Black Family Land Trust (BFLT) to an "African American Land Ethic." Expressing an idea that stems partly from Aldo Leopold's idea of the land ethic, BFLT writes that "land is a tangible asset with economic, human, and spiritual value, which connects African Americans with their rich history in the

Americas and their ancestors. . . . Land ownership represents wealth, power, community, sustainability, and economic opportunities for generations yet born. . . . The history and relationship with the land must be reexamined and self-defined in the context of the African American experience in America. We must utilize every tool available to reduce the rate of African American and other historically underserved populations land loss. The African American Land Ethic is where the BFLT begins: honoring the legacy of those stewards of the land that came before us and having faith in those stewards of the land that will come after us" (Black Family Land Trust, "Land Ethic").

7. hooks, *Belonging*, 201.

8. hooks, 119.

9. hooks, 65 (emphasis added).

10. Berry, *Hidden Wound*, 19. *The Hidden Wound* has influenced many conversations and works on race in the US, including Heather McGhee's acclaimed book *The Sum of Us: What Racism Costs Everyone and How We Can Prosper Together*. In fact, one of McGhee's chapters is titled "The Hidden Wound."

7. Leveraging Love for the Land

1. Jager, *Fate of Family Farming*, 221, 235.

2. Although Jack Sherrick doesn't focus on the virtues of imagination, affection, and fidelity, he thoughtfully describes how some of Berry's agrarian principles and values can be translated into tangible policies ("Prose to Policy").

3. For more on the specifics of the Agricultural Conservation Easement Program and the 2018 Farm Bill, see Land Trust Alliance, "Agricultural Conservation Easement Program."

4. Lehner and Rosenberg, *Farming for Our Future*, 155–156.

5. Strange, *Family Farming*, 83; Ikerd, *Small Farms Are Real Farms*, 13; Bell, *Farming for Us All*.

6. McFadden and Hoppe, "Evolving Distribution of Payments," i; Wickenden, "Wendell Berry's Advice for a Cataclysmic Age"; Burchfield et al., "State of US Farm Operator Livelihoods," 11; Bekkerman, Belasco, and Smith, "Where the Money Goes," 4, 14. In "Growing Farm Size and the Distribution of Farm Payments," James MacDonald, Robert Hoppe, and David Baker emphasize some of these same trends.

7. L. Smith, "EQIP"; Ritter, "Comment"; Bekkerman, Belasco, and Smith, "Where the Money Goes," 14. Recent efforts to direct economic support to underserved groups have been challenged in courts, so policy makers must approach this work with tact.

8. Penniman was quoted in a press release from Senator Cory Booker's office, titled "Booker, Warren, Gillibrand Announce Comprehensive Bill to Address the History of Discrimination in Federal Agricultural Policy."

9. Rappeport, "Black Farmers Fear Foreclosure."

10. USDA, "2017 Census of Agriculture Highlights."

11. Liss, "Policy Update"; Spangler, Burchfield, and Schumacher, "Past and Current Dynamics," 17–18.

12. Lehner and Rosenberg, *Farming for Our Future*, 28–31; Bowlin, "Joke's on Them"; Bonnie, Diamond, and Rowe, "Understanding Rural Attitudes." For an excellent overview of consolidation in the agricultural sector, as well as specific recommendations for addressing this problem, see Kelloway and Miller, "Food and Power"; and Lehner and Rosenberg, *Farming for Our Future*, 28–29.

13. Mitchell, "Restoring Hope for Heirs Property Owners," 9.

14. Jiang and Swallow, "Impact Fees"; Marvier and Wong, "Resurrecting the Conservation Movement." Ballotpedia shows that in a 2012 vote on a constitutional amendment, Alabama residents voted to extend payments to the statewide Forever Wild Land Trust for twenty years. This funding is intended to protect lands, such as recreational areas, nature preserves, and wildlife management areas, from development. The amendment passed overwhelmingly, with over 75 percent of voters choosing to extend funding. Tennessee could consider a similar effort for funding farmland protection. For more information, see Ballotpedia, "Alabama Forever Wild Land Trust Amendment."

15. Two good examples of proactive farmland protection plans can be found in Forsyth County, North Carolina, and Santa Clara County, California. Both counties had the support of their respective states when creating their plans.

16. Valliant et al., "Fostering Farm Transfers." While there are some successful farm-link programs in the South, there are multiple well-known and proven programs located in New England. Through examining these initiatives, one can see the effectiveness of farm-link programs. For more information, see New England Farmland Finder, "New England Farm Link Collaborative."

17. Hulshof, "Tennessee FFA Chapter."

18. See Robertson County, "2040 Comprehensive Growth & Development"; and MACTEC Engineering and Consulting, "Maury County Comprehensive Plan."

19. Cordes, "Agricultural Zoning," 423. In an applied example, a 1981 court case from Illinois upheld that a 160-acre minimum lot size was constitutional and was in accordance with the county government's desire to conserve farmland for public benefit. For more information, see *Wilson v. County of McHenry*, 416 N.E.2d 426 (Ill. App. Ct. 1981). Daniels, "Integrated Working Landscape Protection," 264; US Department of Energy, *Solar Futures Study*; Pedersen and Lamb, "Agrivoltaics."

20. Bengston et al., "Public Policies," 278–279; Farmland Information Center, "Agricultural District Programs." The perk of eminent-domain avoidance could be

especially appealing for folks in Robertson County. For years, there has been talk of expanding Interstate 840—which currently loops around Nashville to the south—through the area. Enrollment in an agricultural district might enable a farmer to prevent condemnation of their land that is otherwise legally unavoidable.

21. John described a "driveway fee" as a small fee levied on any builder who applies for a driveway permit.

22. Daniels, "Integrated Working Landscape Protection," 262.

23. Weible and Sabatier, "Guide to the Advocacy Framework Coalition," 130.

24. Jackson, "Becoming Native to Our Places," 357.

25. Francis et al., "Farmland Conversion," 9.

26. Kotchen and Schulte, "Meta-analysis of Cost of Community Services Studies," 377.

27. Over an eighteen-year period, American Farmland Trust completed 128 CCS studies in states across the nation. In 127 of these studies, agricultural land and open space generated better ratios of revenue-to-expenses than residential land uses. In Robertson County, commercial and industrial land uses also brought in more revenue than they required in expense. American Farmland Trust, "Cost of Community Services Study," 2, 17.

28. Gallagher, "Environmental, Social, and Cultural Impacts," 219, 267; J. White, "Beating Plowshares into Townhomes," 115.

29. Sims et al., "Economic Value of Open Space," 5.

30. These ecosystem services include flood mitigation, wildlife habitat, pollination, air-pollution removal, carbon sequestration, and more. Sims et al.

31. American Farmland Trust, *Greener Fields: California*; American Farmland Trust, *Greener Fields: New York*.

32. Szlanfucht, "How to Save America's Depleting Supply," 340; Johnson and Klemens, "Impacts of Sprawl on Biodiversity," 103–106; Lehner and Rosenberg, *Farming for Our Future*, 164; Groh, Isenhart, and Schultz, "Long-Term Nitrate Removal"; Lee, Isenhart, and Schultz, "Sediment and Nutrient Removal."

33. Bengston et al., "Analysis of the Public Discourse," 746; Johnson and Klemens, "Impacts of Sprawl on Biodiversity," 19–22; Gallagher, "Environmental, Social, and Cultural Impacts," 221–222.

34. Sorenson and Hunter, *Wildlife on the Working Landscape*; TWRA, *Tennessee State Wildlife Action Plan*.

35. For a specific example of the importance of virtuous hope in giving rise to a better future for farming, see Graddy, "Legal and Legislative Front," 233.

Bibliography

Adams, Susan, and Hank Tucker. "For HBCUs Cheated out of Billions, Bomb Threats Are the Latest Indignity." *Forbes*, February 1, 2022. https://www.forbes.com/sites/susanadams/2022/02/01/for-hbcus-cheated-out-of-billions-bomb-threats-are-latest-indignity.

Agee, James, and Walker Evans. *Let Us Now Praise Famous Men.* 1939. Reprint, Boston: Houghton Mifflin, 2001.

Albritton, Lesley, and Jesse Williams. "Disasters Do Discriminate: Black Land Tenure and Disaster Relief Programs." *Journal of Affordable Housing* 29, no. 3 (2021): 421–447.

American Farmland Trust. "Cost of Community Services Study: Robertson County, Tennessee." Washington, DC, June 2006.

———. *Greener Fields: California Communities Combating Climate Change.* Washington, DC, August 2018. https://farmlandinfo.org/publications/greener-fields-california-communities-combating-climate-change/.

———. *Greener Fields: Combating Climate Change by Keeping Land in Farming in New York.* Washington, DC, May 2017. https://farmland-info.org/publications/greener-fields-combating-climate-change-by-keeping-land-in-farming-in-new-york/.

———. "Tennessee: Agricultural Land Conversion Highlight Summary." Washington, DC, 2020.

Annas, Julia. *Intelligent Virtue.* Oxford: Oxford University Press, 2011.

Aristotle. *Nicomachean Ethics.* Translated by Robert C. Bartlett and Susan D. Collins. Chicago: University of Chicago Press, 2011.

Bai, Heesoon, David Chang, and Charles Scott, eds. *A Book of Ecological Virtues: Living Well in the Anthropocene.* Regina, SK: University of Regina Press, 2020.

Bailey, Conner, and Ryan Thomson. "Heirs Property, Critical Race Theory, and Reparations." *Rural Sociology*, June 17, 2022.

Bailey, Liberty Hyde. *The Holy Earth*. 1915. Reprint, Berkeley, CA: Counterpoint, 2015.

Baker, Andrew C. *Bulldozer Revolutions: A Rural History of the Metropolitan South*. Athens: University of Georgia Press, 2018.

Ballotpedia. "Alabama Forever Wild Land Trust Amendment, Amendment 1 (2012)." Accessed July 25, 2022. https://ballotpedia.org/Alabama_Forever_Wild_Land_Trust_Amendment,_Amendment_1_(2012).

Balthrop-Lewis, Alda. "Exemplarist Environmental Ethics: Thoreau's Political Ascetism against Solution Thinking." *Journal of Religious Ethics* 47, no. 3 (September 2019): 525–550.

Balvanz, Peter, Morgan L. Barlow, Lillianne M. Lewis, Kari Samuel, William Owens, Donna L. Parker, Molly De Marco, Robin Crowder, Yarbrough Williams, Dorathy Barker, Alexandra Lightfoot, and Alice Ammerman. " 'The Next Generation, That's Why We Continue to Do What We Do': African American Farmers Speak about Experiences with Land Ownership and Loss in North Carolina." *Journal of Agriculture, Food Systems, and Community Development* 1, no. 3 (2011): 67–88.

Barnett, Barry. "The U.S. Farm Financial Crisis of the 1980s." *Agricultural History* 74, no. 2 (2000): 366–380.

Bekkerman, Anton, Eric J. Belasco, and Vincent H. Smith. "Where the Money Goes: The Distribution of Crop Insurance and Other Farm Subsidy Payments." In *Agricultural Policy in Disarray: Reforming the Farm Bill*. Washington, DC: American Enterprise Institute, January 2018.

Bell, Michael M. *Farming for Us All: Practical Agriculture and the Cultivation of Sustainability*. University Park: Pennsylvania State University Press, 2004.

Bengston, David N., Jennifer O. Fletcher, and Kristen C. Nelson. "Public Policies for Managing Urban Growth and Protecting Open Space: Policy Instruments and Lessons Learned in the United States." *Landscape and Urban Planning* 69 (2004): 271–286.

Bengston, David N., Robert S. Potts, David P. Fan, and Edward G. Goetz. "An Analysis of the Public Discourse about Urban Sprawl in the United

States: Monitoring Concern about a Major Threat to Forests." *Forest Policy and Economics* 7 (2005): 745–756.

Berry, Wendell. *The Art of Loading Brush: New Agrarian Writings.* Berkeley, CA: Counterpoint, 2017.

———. "The Body and the Earth." In *The Unsettling of America: Culture and Agriculture,* 101–145. 1977. Reprint, Berkeley, CA: Counterpoint, 2015.

———. "Conservationist and Agrarian." 2002. In *Bringing It to the Table: On Farming and Food,* 67–79. Berkeley, CA: Counterpoint, 2009.

———. "A Defense of the Family Farm." 1986. In *Bringing It to the Table: On Farming and Food,* 31–48. Berkeley, CA: Counterpoint, 2009.

———. "Farmland without Farmers." *Atlantic,* March 19, 2015. https://www.theatlantic.com/national/archive/2015/03/farmland-without-farmers/388282/.

———. *The Gift of Good Land.* Berkeley, CA: Counterpoint, 1981.

———. "Health Is Membership." In *Another Turn of the Crank,* 86–109. Berkeley, CA: Counterpoint, 1995.

———. *The Hidden Wound.* 1970. Reprint, Berkeley, CA: Counterpoint, 2010.

———. "Imagination in Place." 2004. In *Imagination in Place,* 1–16. Berkeley, CA: Counterpoint, 2010.

———. "It All Turns on Affection." In *It All Turns on Affection: The Jefferson Lecture and Other Essays,* 9–39. Berkeley, CA: Counterpoint, 2012.

———. "A Native Hill." 1969. In *The Art of the Commonplace: The Agrarian Essays of Wendell Berry,* edited by Norman Wirzba, 3–31. Berkeley, CA: Counterpoint, 2002.

———. "Notes from an Absence and a Return." 1970. In *A Continuous Harmony,* 35–53. Berkeley, CA: Counterpoint, 2012.

———. "People, Land, and Community." 1983. In *The Art of the Commonplace: The Agrarian Essays of Wendell Berry,* edited by Norman Wirzba, 182–194. Berkeley, CA: Counterpoint, 2002.

———. "A Remarkable Man." 1975. In *What Are People For?,* 17–29. Berkeley, CA: Counterpoint, 2010.

———. "Renewing Husbandry." 2004. In *Bringing It to the Table: On Farming and Food,* 87–101. Berkeley, CA: Counterpoint, 2009.

————. "Stand by Me." In *A Place in Time: Twenty Stories of the Port William Membership*, 98–112. Berkeley, CA: Counterpoint, 2012.

————. "Starting from Loss." In *It All Turns on Affection: The Jefferson Lecture and Other Essays*, 67–88. Berkeley, CA: Counterpoint, 2012.

————. "Think Little." 1970. In *The Art of the Commonplace: The Agrarian Essays of Wendell Berry*, edited by Norman Wirzba, 81–90. Berkeley, CA: Counterpoint, 2002.

————. *The Unsettling of America: Culture and Agriculture*. 1977. Reprint, Berkeley, CA: Counterpoint, 2015.

Berry, Wendell, and Wes Jackson. "A 50-Year Farm Bill." *New York Times*, January 4, 2009. https://www.nytimes.com/2009/01/05/opinion/05berry.html.

Bilbro, Jeffrey. *Virtues of Renewal: Wendell Berry's Sustainable Forms*. Lexington: University Press of Kentucky, 2019.

Black Family Land Trust. "Land Ethic." Accessed February 18, 2022. https://www.bflt.org/land-ethic.html.

Blum, Elizabeth D. "Power, Danger, and Control: Slave Women's Perceptions of Wilderness in the Nineteenth Century." *Women's Studies* 31, no. 2 (2002): 247–265.

Bonnie, Robert, Emily Pechar Diamond, and Elizabeth Rowe. "Understanding Rural Attitudes toward the Environment and Conservation in America." Nicholas Institute for Environmental Policy Solutions, Duke University, 2020.

Booker, Cory. "Booker, Warren, Gillibrand Announce Comprehensive Bill to Address the History of Discrimination in Federal Agricultural Policy." Press release, November 19, 2020. https://www.booker.senate.gov/news/press/-booker-warren-gillibrand-announce-comprehensive-bill-to-address-the-history-of-discrimination-in-federal-agricultural-policy.

Bowlin, Nick. "Joke's on Them: How Democrats Gave Up on Rural America." *Guardian*, February 22, 2022. https://www.theguardian.com/us-news/2022/feb/22/us-politics-rural-america.

Brown, David, and Kai Schafft. *Rural People and Communities in the 21st Century: Resilience and Transformation*. Malden, MA: Polity, 2011.

Bunge, Jacob. "Supersized Family Farms Are Gobbling Up American Agriculture." *Wall Street Journal*, October 23, 2017. https://www.wsj.com/articles/the-family-farm-bulks-up-1508781895.

Burchfield, Emily K., Britta L. Schumacher, Kaitlyn Spangler, and Andrea Rissing. "The State of US Farm Operator Livelihoods." *Frontiers in Sustainable Food Systems* 5 (2022): 1–22.

Cafaro, Philip. "Thoreau, Leopold, and Carson: Toward an Environmental Virtue Ethic." *Environmental Ethics* 23, no. 1 (Spring 2001): 3–17.

Canaday, Neil, and Charles Reback. "Race, Literacy, and Real Estate Transactions in the Postbellum South." *Journal of Economic History* 70, no. 2 (2010): 428–445.

Carney, Judith A. *Black Rice: The African Origins of Rice Cultivation in the Americas*. Cambridge, MA: Harvard University Press, 2001.

Carr, Donald, and Anne Schechinger. "USDA Pandemic Bailout Funds Will Go to Largest, Wealthiest Farms." Environmental Working Group, May 21, 2020. https://www.ewg.org/news-insights/news/usda-pandemic-bailout-funds-will-go-largest-wealthiest-farms.

Castro, Abril, and Zoe Willingham. "Progressive Governance Can Turn the Tide for Black Farmers." Center for American Progress, April 3, 2019. https://www.americanprogress.org/issues/economy/reports/2019/04/03/467892/progressive-governance-can-turn-tide-black-farmers/.

Charles, Safiya. "Debt, Racism, and Fear of Displacement Are Driving an Overlooked Public Health Crisis among Black Farmers." *The Counter*, March 17, 2022. https://thecounter.org/black-farmers-racism-public-health-research.

Conklin, Paul K. *A Revolution Down on the Farm: The Transformation of American Agriculture since 1929*. Lexington: University Press of Kentucky, 2008.

Cordes, Mark W. "Agricultural Zoning: Impacts and Future Directions." *Northern Illinois University Law Review* 22, no. 3 (2002): 419–458.

Coulthard, Robert A. "The Changing Landscape of America's Farmland: A Comparative Look at Policies Which Help Determine the Portrait of Our Land—Are There Lessons We Can Learn from the EU?" *Drake Journal of Agricultural Law* 6, no. 2 (Fall 2001): 261–286.

Cuomo, Christine J. "Respect for Nature: Learning from Indigenous Values." In *The Virtues of Sustainability*, edited by Jason Kawall, 135–157. Oxford: Oxford University Press, 2021.

Daniel, Pete. "African American Farmers and Civil Rights." *Journal of Southern History* 73, no. 1 (2007): 3–38.

———. *Dispossession: Discrimination against African American Farmers in the Age of Civil Rights*. Chapel Hill: University of North Carolina Press, 2013.

Daniels, Tom. "Integrated Working Landscape Protection: The Case of Lancaster County, Pennsylvania." *Society & Natural Resources* 13, no. 3 (2000): 261–271.

———. *When City and Country Collide: Managing Growth in the Metropolitan Fringe*. Washington, DC: Island, 1999.

Darity, William, Jr. "Forty Acres and a Mule in the 21st Century." *Social Science Quarterly* 89, no. 3 (2008): 657–664.

Davis, Ellen F. *Scripture, Culture, and Agriculture: An Agrarian Reading of the Bible*. New York: Cambridge University Press, 2009.

DiEnno, Cara Marie, and Jessica Leigh Thompson. "For the Love of the Land: How Emotions Motivate Volunteerism in Ecological Restoration." *Emotion, Space and Society* 6 (2013): 63–72.

Dodson Gray, Elizabeth. "A Critique of Dominion Theology." In *For Creation's Sake: Preaching, Ecology, and Justice*, edited by Dieter Hessel, 71–83. Philadelphia: Geneva, 1985.

Donahue, Brian. *Go Farm, Young People, and Help Heal the Country*. Common Sense for Global Crises—A Pamphlet Series. Middlebury, VT: New Perennials, 2022.

Dudley, Kathryn. *Debt and Dispossession: Farm Loss in America's Heartland*. Chicago: University of Chicago Press, 2000.

Dunlap, Riley E. "Conventional versus Alternative Agriculture: The Paradigmatic Roots of the Debate." *Rural Sociology* 55, no. 4 (December 1990): 590–616.

Dyer, Janice F., and Conner Bailey. "A Place to Call Home: Cultural Understandings of Heir Property among Rural African Americans." *Rural Sociology* 73, no. 3 (2008): 317–338.

Equal Justice Initiative. "Community Remembrance Project." Accessed February 16, 2022. https://eji.org/projects/community-remembrance-project/.

Ewing, Reid, and Shima Hamidi. *Measuring Sprawl 2014*. Washington, DC: Smart Growth America, 2014.

Fairbairn, Madeleine. *Fields of Gold: Financing the Global Land Rush*. Ithaca, NY: Cornell University Press, 2020.

Farmland Information Center, American Farmland Trust. "Agricultural Districts Program." Washington, DC, 2016. https://farmlandinfo.org/publications/agricultural-district-programs/.

Farrell, Justin. *The Battle for Yellowstone: Morality and the Sacred Roots of Environmental Conflict*. Princeton, NJ: Princeton University Press, 2015.

Ferdman, Roberto A. "The Decline of the Small American Family Farm in One Chart." *Washington Post*, September 16, 2014. https://www.washingtonpost.com/news/wonk/wp/2014/09/16/the-decline-of-the-small-american-family-farm-in-one-chart/.

Ferrell, John S. "George Washington Carver: A Blazer of Trails to a Sustainable Future." In *Land and Power: Sustainable Agriculture and African Americans*, edited by Jeffrey L. Jordan, Edward Pennick, Walter A. Hill, and Robert Zabawa, 11–32. Sustainable Agriculture and Research (SARE) Program. Waldorf, MD: Sustainable Agriculture Publications, 2009.

FFA (Future Farmers of America). "FFA Creed." Accessed July 25, 2022. https://www.ffa.org/about/ffa-creed/.

Finney, Carolyn. *Black Faces, White Spaces: Reimagining the Relationship of African Americans to the Great Outdoors*. Chapel Hill: University of North Carolina Press, 2014.

Francis, Charles A., Twyla E. Hansen, Allison A. Fox, Paula J. Hesje, Hana E. Nelson, Andrea E. Lawseth, and Alexandra English. "Farmland Conversion to Non-agricultural Uses in the US and Canada: Current Impacts and Concerns for the Future." *International Journal of Agricultural Sustainability* 10, no. 1 (February 2012): 8–24.

Francis, Dania V., Darrick Hamilton, Thomas W. Mitchell, Nathan A. Rosenberg, and Bryce Wilson Stucki. "Black Land Loss: 1920–1997." *American Economic Association Papers and Proceedings* 112 (May 2022): 38–42.

Francis, Greg. *Just Harvest: The Story of How Black Farmers Won the Largest Civil Rights Case against the U.S. Government*. Brentwood, TN: Forefront Books, 2021.

Freedgood, Julia, Mitch Hunter, Jennifer Dempsey, and Ann Sorensen. *Farms under Threat: The State of the States.* Washington, DC: American Farmland Trust, 2020.

Freilich, Robert, Robert J. Sitkowski, and Seth D. Mennillo. *From Sprawl to Sustainability: Smart Growth, New Urbanism, Green Development, and Renewable Energy.* 2nd ed. Chicago: American Bar Association, 2010.

Gallagher, Patrick. "The Environmental, Social, and Cultural Impacts of Sprawl." *Natural Resources & Environment* 15, no. 4 (2001): 219–223, 267.

Gilbert, Jess, Gwen Sharp, and M. Sindy Felin. "The Loss and Persistence of Black-Owned Farms and Farmland: A Review of the Research Literature and Its Implications." *Southern Rural Sociology* 18, no. 2 (2002): 1–30.

Glymph, Thavolia. *The Women's Fight: The Civil War's Battles for Home, Freedom, and Nation.* Chapel Hill: University of North Carolina Press, 2020.

Goldschmidt, Walter. *As You Sow: Three Studies in the Social Consequences of Agribusiness.* New York: Harcourt Brace, 1947.

Goodwin, Joy N., and Jessica L. Gouldthorpe. "Small Farmers, Big Challenges: A Needs Assessment of Florida Small-Scale Farmers' Production Challenges and Training Needs." *Journal of Rural Social Sciences* 28, no. 1 (2013): 54–79.

Graddy, Hank. "The Legal and Legislative Front: The Fight against Industrial Agriculture." In *The Essential Agrarian Reader: The Future of Culture, Community, and the Land,* edited by Norman Wirzba, 222–236. Berkeley, CA: Counterpoint, 2004.

Grim, Valerie. "Between Forty Acres and a Class Action Lawsuit: Black Farmers, Civil Rights, and Protest against the U.S. Department of Agriculture." In *Beyond Forty Acres and a Mule: African American Landowning Families since Reconstruction,* edited by Debra A. Reid and Evan P. Bennett, 271–296. Gainesville: University Press of Florida, 2012.

Groh, Tyler A., Thomas M. Isenhart, and Richard C. Schultz. "Long-Term Nitrate Removal in Three Riparian Buffers: 21 Years of Data from the Bear Creek Watershed in Central Iowa, USA." *Science of the Total Environment* 740 (2020): 1–12.

Hanson, J. D., John Hendrickson, and Dave Archer. "Challenges for Maintaining Sustainable Agricultural Systems in the United States." *Renewable Agriculture and Food Systems* 23, no. 4 (2008): 325–334.

Hinson, Waymon R., and Edward Robinson. " 'We Didn't Get Nothing': The Plight of Black Farmers." *Journal of African American Studies* 12 (2008): 283–302.

hooks, bell. *All about Love: New Visions.* New York: William Morrow, 2001.

———. *Belonging: A Culture of Place.* New York: Routledge, 2009.

Horst, Megan, and Amy Marlon. "Racial, Ethnic and Gender Inequities in Farmland Ownership and Farming in the U.S." *Agriculture and Human Values* 36 (2019): 1–16.

Hribar, Carrie. *Understanding Concentrated Animal Feeding Operations and Their Impact on Communities.* Bowling Green, OH: National Association of Local Boards of Health, 2010. https://www.cdc.gov/nceh/ehs/docs/understanding_cafos_nalboh.pdf.

Huebner, Timothy S. *Liberty and Union: The Civil War Era and American Constitutionalism.* Lawrence: University Press of Kansas, 2016.

Hulshof, Kacie. "Tennessee FFA Chapter Explores Black-Rooted NFA History." *AgDaily*, 2021. https://www.agdaily.com/ffa/tennessee-ffa-chapter-explores-nfa-history.

Humphrey, Cam. "Centering a Population Overlooked in Agriculture: The Young Black Farmer." *Yale Environment Review*, November 25, 2020. https://environment-review.yale.edu/centering-population-overlooked-agriculture-young-black-farmer.

Hunter, Mitch, Ann Sorenson, Theresa Nogeire-McRae, Scott Beck, Stacy Shutts, and Ryan Murphy. *Farms under Threat 2040: Choosing an Abundant Future.* Washington, DC: American Farmland Trust, 2022.

Ikerd, John. *Small Farms Are Real Farms: Sustaining People through Agriculture.* Austin, TX: Acres USA, 2008.

Irwin, Elena G., Hyun Jin Cho, and Nancy E. Bockstael. "Measuring the Amount and Pattern of Land Development in Nonurban Areas." *Review of Agricultural Economics* 29, no. 3 (2007): 494–501.

Jackson, Wes. "Becoming Native to Our Places." 1994. In *American Georgics: Writings on Farming, Culture, and the Land*, edited by Edwin C. Hagenstein,

Sara M. Gregg, and Brian Donahue, 347–358. New Haven, CT: Yale University Press, 2011.

———. "Prologue." In *Becoming Native to This Place*, 1–5. Lexington: University Press of Kentucky, 1994.

Jager, Ronald. *The Fate of Family Farming: Variations on an American Idea.* Lebanon, NH: University Press of New England, 2004.

Jeong, Yihyun. "Nashville Approves New Budget with 34 Percent Tax Hike, More Funds for Police, Schools." *Nashville Tennessean*, June 17, 2020. https://www.tennessean.com/story/news/politics/2020/06/17/nashville-approves-new-budget-34-percent-tax-hike-increase-funds-police-schools/3200993001/.

Jiang, Yong, and Stephen K. Swallow. "Impact Fees Coupled with Conservation Payments to Sustain Ecosystem Structure: A Conceptual and Numerical Application at the Urban-Rural Fringe." *Ecological Economics* 136 (2017): 136–147.

Johnson, Elisabeth A., and Michael W. Klemens. "The Impacts of Sprawl on Biodiversity." In *Nature in Fragments: The Legacy of Sprawl*, edited by Elisabeth A. Johnson and Michael W. Klemens, 18–53. New York: Columbia University Press, 2005.

Johnson, Roger. "We Must Reject the 'Go Big or Go Home' Mentality of Modern Agriculture." *The Hill*, October 8, 2019. https://thehill.com/opinion/finance/464856-we-must-reject-the-go-big-or-go-home-mentality-of-modern-agriculture.

Jones, Allen. "Thomas M. Campbell: Black Agricultural Leader of the New South." *Agricultural History* 53, no. 1 (1979): 42–59.

Kawall, Jason, ed. *The Virtues of Sustainability*. Oxford: Oxford University Press, 2021.

Kelloway, Claire, and Sarah Miller. "Food and Power: Addressing Monopolization in America's Food System." Open Markets Institute, March 2019.

Kimbrell, Andrew, ed. *The Fatal Harvest Reader: The Tragedy of Industrial Agriculture*. Sausalito, CA: Foundation for Deep Ecology, 2002.

Kingsolver, Barbara. Foreword to *The Essential Agrarian Reader: The Future of Culture, Community, and the Land*, edited by Norman Wirzba, ix–xvii. Berkeley, CA: Counterpoint, 2004.

Koerth, Maggie. "Big Farms Are Getting Bigger and Most Small Farms Aren't Really Farms at All." *FiveThirtyEight*, November 17, 2016. https://fivethirtyeight.com/features/big-farms-are-getting-bigger-and-most-small-farms-arent-really-farms-at-all/.

Kotchen, Matthew J., and Stacey L. Schulte. "A Meta-analysis of Cost of Community Services Studies." *International Regional Science Review* 32, no. 3 (2009): 376–399.

Kremen, Claire, Alastair Iles, and Christopher Bacon. "Diversified Farming Systems: An Agroecological, Systems-Based Alternative to Modern Industrial Agriculture." *Ecology and Society* 17, no. 4 (December 2012): 44–63.

Kremer, Gary R., ed. *George Washington Carver: In His Own Words.* 2nd ed. Columbia: University of Missouri Press, 2017.

Lamb, Brooks. *Overton Park: A People's History.* Knoxville: University of Tennessee Press, 2019.

———. "Understanding Heirs' Property and Its Impact on Farmers." *AgDaily*, February 23, 2022. https://www.agdaily.com/insights/understanding-heirs-property-and-its-impact-on-farmers/.

Lamb, Michael. "Difficult Hope: Wendell Berry and Climate Change." In *Hope*, edited by Nancy E. Snow. Oxford: Oxford University Press, forthcoming.

Land Trust Alliance. "Agricultural Conservation Easement Program." Accessed December 8, 2022. https://landtrustalliance.org/resources/learn/explore/ale-toolkit/.

———. "Frequently Asked Questions." Accessed December 8, 2022. https://landtrustalliance.org/take-action/conserve-your-land/frequently-asked-questions.

Lanham, J. Drew. *The Home Place: Memoirs of a Colored Man's Love Affair with Nature.* Minneapolis: Milkweed Editions, 2016.

Larmer, Megan. "Cultivating the Edge: An Ethnography of First-Generation Women Farmers in the American Midwest." *Feminist Review* 114 (2016): 91–111.

Lee, Jung. "Ethnography and Ethics." In *International Encyclopedia of Ethics*, edited by Hugh LaFollette, 1–11. Hoboken, NJ: Wiley, 2017. https://doi.org/10.1002/9781444367072.wbiee835.

Lee, Kye-Han, Thomas M. Isenhart, and Richard C. Schultz. "Sediment and Nutrient Removal in an Established Multi-species Riparian Buffer." *Journal of Soil and Water Conservation* 58, no. 1 (2003): 1–8.

Lehner, Peter H., and Nathan A. Rosenberg. *Farming for Our Future: The Science, Law, and Policy of Climate-Neutral Agriculture.* Washington, DC: Environmental Law Institute, 2021.

Leopold, Aldo. *A Sand County Almanac.* 1949. Reprint, New York: Ballantine Books, 1970.

LeVasseur, Todd. *Religious Agrarianism and the Return of Place: From Values to Practice in Sustainable Agriculture.* Albany: State University of New York Press, 2017.

Liss, Emily. "Policy Update: AFT Submits Comment to USDA on Racial Equity." American Farmland Trust, August 4, 2021. https://farmland.org/policy-update-aft-submits-comment-to-usda-on-racial-equity/.

Littlefield, Daniel C. *Rice and Slaves: Ethnicity and the Slave Trade in South Carolina.* Urbana: University of Illinois Press, 1991.

Lobao, Linda M. *Locality and Inequality: Farm and Industry Structure and Socioeconomic Conditions.* Albany: State University of New York Press, 1990.

Lyson, Thomas A., Robert J. Torres, and Rick Welsh. "Scale of Agricultural Production, Civic Engagement, and Community Welfare." *Social Forces* 80, no. 1 (2001): 311–327.

MacDonald, James, Robert Hoppe, and David Banker. "Growing Farm Size and the Distribution of Farm Payments." Economic Brief 6. US Department of Agriculture, Economic Research Service, March 2006.

MacDonald, James, Robert Hoppe, and Doris Newton. "Three Decades of Consolidation in U.S. Agriculture." Economic Information Bulletin 189. US Department of Agriculture, Economic Research Service, March 2018.

MACTEC Engineering and Consulting. "Maury County Comprehensive Plan." August 1, 2011. https://www.maurycounty-tn.gov/DocumentCenter/View/236/Maury-County-Comprehensive-Plan-PDF.

Marty, Martin E. "The Terror of Land Loss, the Dream of Finding Home." In *The Longing for Home,* edited by Leroy S. Rouner, 243–263. Notre Dame, IN: University of Notre Dame Press, 1996.

Marvier, Michelle, and Hazel Wong. "Resurrecting the Conservation Movement." *Journal of Environmental Studies and Sciences* 2 (2012): 291–295.

McFadden, Jonathan R., and Robert A. Hoppe. "The Evolving Distribution of Payments from Commodity, Conservation, and Federal Crop Insurance Programs." Economic Information Bulletin 184. US Department of Agriculture, Economic Research Service, November 2017.

McGhee, Heather. *The Sum of Us: What Racism Costs Everyone and How We Can Prosper Together.* New York: One World, 2021.

Miller, Melinda C. "Land and Racial Wealth Inequality." *American Economic Review* 101, no. 3 (2011): 371–376.

Minor, Kelly A. " 'Justifiable Pride': Negotiation and Collaboration in Florida African American Extension." In *Beyond Forty Acres and a Mule: African American Landowning Families since Reconstruction*, edited by Debra A. Reid and Evan P. Bennett, 205–228. Gainesville: University Press of Florida, 2012.

Mitchell, Thomas W. "Restoring Hope for Heirs Property Owners: The Uniform Partition of Heirs Property Act." Texas A&M University School of Law Legal Studies Research Paper No. 17-04. *State & Local Law News* 40, no. 1 (2016): 6–15.

Moran, Greta. "Beginning Farmers, Farmers of Color Outbid as Farmland Prices Soar." *Civil Eats*, January 3, 2022. https://civileats.com/2022/01/03/beginning-farmers-farmers-of-color-outbid-as-farmland-prices-soar/.

Moroney, Jillian I., and Rebecca Som Castellano. "Farmland Loss and Concern in the Treasure Valley." *Agriculture and Human Values* 35 (2018): 529–536.

National Endowment for the Humanities. "2012 Jefferson Lecture with Wendell Berry." April 25, 2012. https://www.neh.gov/news/2012-jefferson-lecture-wendell-berry.

New England Farmland Finder. "New England Farm Link Collaborative." Accessed July 26, 2022. https://newenglandfarmlandfinder.org/new-england-farm-link-collaborative.

NPR (National Public Radio). " 'Theft at a Scale That Is Unprecedented': Behind the Underfunding of HBCUs." May 13, 2021. https://www.npr.org/2021/05/13/996617532/behind-the-underfunding-of-hbcus.

Oehlschlaeger, Fritz. *The Achievement of Wendell Berry: The Hard History of Love*. Lexington: University of Kentucky Press, 2011.

Olmstead, Grace. *Uprooted: Recovering the Legacy of the Places We've Left Behind*. New York: Sentinel, 2021.

Olson, Richard, and Thomas Lyson. *Under the Blade: The Conversion of Agricultural Landscapes*. Boulder, CO: Westview, 1999.

Parvini, Sarah. "Nashville's Southern Hospitality—and Affordability—Beckon Californians." *Los Angeles Times,* December 28, 2021. https://www.latimes.com/california/story/2021-12-28/californians-moving-nashville.

Pedersen, Bill, and Brooks Lamb. "Agrivoltaics: Producing Solar Energy While Protecting Farmland." Center for Business and the Environment at Yale, October 2021. https://cbey.yale.edu/research/agrivoltaics-producing-solar-energy-while-protecting-farmland.

Pennick, Edward Jerry, Heather Gray, and Miessha N. Thomas. "Preserving African American Rural Property: An Assessment of Intergenerational Values Toward Land." In *Land and Power: Sustainable Agriculture and African Americans*, edited by Jeffrey L. Jordan, Edward Pennick, Walter A. Hill, and Robert Zabawa, 153–173. Sustainable Agriculture and Research (SARE) Program. Waldorf, MD: Sustainable Agriculture Publications, 2009.

Penniman, Leah. *Farming While Black: Soul Fire Farm's Practical Guide to Liberation on the Land*. White River Junction, VT: Chelsea Green, 2018.

Peters, Jason, ed. *Wendell Berry: Life and Work*. Lexington: University Press of Kentucky, 2007.

Petrusich, Amanda. "Going Home with Wendell Berry." *New Yorker*, July 14, 2019. https://www.newyorker.com/culture/the-new-yorker-interview/going-home-with-wendell-berry.

Petty, Adrienne. "The Jim Crow Section of Agricultural History." In *Beyond Forty Acres and a Mule: African American Landowning Families since Reconstruction*, edited by Debra A. Reid and Evan P. Bennett, 21–35. Gainesville: University Press of Florida, 2012.

Pilgeram, Ryanne. *Pushed Out: Contested Development and Rural Gentrification in the U.S. West*. Seattle: University of Washington Press, 2021.

Presser, Lizzie. "Their Family Bought Land One Generation after Slavery: The Reels Brothers Spent Eight Years in Jail for Refusing to Leave It." ProPublica, July 15, 2019. https://features.propublica.org/black-land-loss/heirs-property-rights-why-black-families-lose-land-south/.

Preston, Jane. "Community Involvement in School: Social Relationships in a Bedroom Community." *Canadian Journal of Education* 36, no. 3 (2013): 413–437.

Rappeport, Alan. "Black Farmers Fear Foreclosure as Debt Relief Remains Frozen." *New York Times*, February 21, 2022. https://www.nytimes.com/2022/02/21/us/politics/black-farmers-debt-relief.html.

Rebanks, James. *Pastoral Song: A Farmer's Journey*. New York: HarperCollins—Custom House, 2020.

Reid, Debra A., and Evan P. Bennett, eds. *Beyond Forty Acres and a Mule: African American Landowning Families since Reconstruction*. Gainesville: University Press of Florida, 2012.

Reynolds, Bruce J. *Black Farmers in the Pursuit of Independent Farming and the Role of Cooperatives*. Washington, DC: USDA Rural Business Cooperative, 2002.

Ritter, Tara. "Comment on the Environmental Quality Incentives Program Interim Final Rule." Institute for Agriculture and Trade Policy, February 19, 2020. https://www.iatp.org/documents/comment-environmental-quality-incentives-program-interim-final-rule.

Roberts, Robert. "Will Power and the Virtues." *Philosophical Review* 93, no. 2 (April 1984): 227–247.

Robertson County. "2040 Comprehensive Growth & Development." September 2013. https://robertsoncountytn.gov/departments/planning/2040_comprehensive_growth___development.php.

Rome, Adam. *The Bulldozer in the Countryside: Suburban Sprawl and the Rise of American Environmentalism*. New York: Cambridge University Press, 2001.

Rosengarten, Theodore. *All God's Dangers: The Life of Nate Shaw*. 1974. Reprint, New York: Vintage Books, 2018.

Ruffin, Kimberly N. *Black on Earth: African American Ecoliterary Traditions*. Athens: University of Georgia Press, 2010.

Rufus, Musonius. "Lecture No. 11." In *How to Be a Farmer: An Ancient Guide to Life on the Land*, translated by M. D. Usher, 183–195. Princeton, NJ: Princeton University Press, 2021.

Salamon, Sonya. "From Hometown to Nontown: Rural Community Effects of Suburbanization." *Rural Sociology* 68, no. 1 (2003): 1–24.

———. *Newcomers to Old Towns: Suburbanization of the Heartland*. Chicago: University of Chicago Press, 2003.

Sandler, Ronald L. *Character and Environment: A Virtue-Oriented Approach to Environmental Ethics*. New York: Columbia University Press, 2007.

Schechter, Patricia A. "On Violence in the South: Ida B. Wells-Barnett." Center for the Study of Southern Culture, University of Mississippi, July 11, 2016. https://southernstudies.olemiss.edu/on-violence-in-the-south-ida-b-wells-barnett/.

Schultz, Mark. "Benjamin Hubert and the Association for the Advancement of Negro Country Life." In *Beyond Forty Acres and a Mule: African American Landowning Families since Reconstruction*, edited by Debra A. Reid and Evan P. Bennett, 83–105. Gainesville: University Press of Florida, 2012.

———. *The Rural Face of White Supremacy: Beyond Jim Crow*. Urbana: University of Illinois Press, 2005.

Scott, James C. *Seeing Like a State: How Certain Schemes to Improve the Human Condition Have Failed*. New Haven, CT: Yale University Press, 1999.

Semuels, Alana. " 'They're Trying to Wipe Us Off the Map': Small American Farmers Are Nearing Extinction." *Time*, November 27, 2019. https://time.com/5736789/small-american-farmers-debt-crisis-extinction/.

Sewell, Summer. "There Were Nearly a Million Black Farmers in 1920. Why Have They Disappeared?" *Guardian*, April 29, 2019. https://www.theguardian.com/environment/2019/apr/29/why-have-americas-black-farmers-disappeared.

Sherman, Jennifer. *Dividing Paradise: Rural Inequality and the Diminishing American Dream*. Oakland: University of California Press, 2021.

Sherrick, Jack. "Prose to Policy: How Wendell Berry's Distinct Strain of Agrarianism Can Influence Farm Policy." *Columbia Journal of Law and Social Problems* 56, no. 2 (forthcoming).

Sherval, Meg, Hedda Haugen Askland, Michael Askew, Jo Hanley, David Farrugia, Steven Threadgold, and Julia Coffey. "Farmers as Modern-Day Stewards and the Rise of New Rural Citizenship in the Battle over Land Use." *Local Environment: The International Journal of Justice and Sustainability* 23, no. 1 (2018): 100–116.

Sims, Charles, Jill Welch, Rebecca J. Davis, Yinan Liu, Doug Yan, Cassidy Quistorff, and Matthew N. Murray. "The Economic Value of Open Space in the Cumberland Region." Howard H. Baker Jr. Center for Public Policy, University of Tennessee, May 7, 2018.

Smith, John David, ed. *Black Soldiers in Blue: African American Troops in the Civil War Era.* Chapel Hill: University of North Carolina Press, 2004.

Smith, Kimberly K. *Wendell Berry and the Agrarian Tradition: A Common Grace.* Lawrence: University Press of Kansas, 2003.

Smith, Lexi. "EQIP: Subsidies for Big Ag in the Farm Bill's Conservation Title." Center for Health, Law, and Policy Innovation, Harvard Law School, September 7, 2017. https://chlpi.org/news-and-events/news-and-commentary/commentary/eqip-subsidies-big-ag-farm-bills-conservation-title/.

Sorenson, Ann, and Mitch Hunter. *Wildlife on the Working Landscape: Charting a Way for Biodiversity and Agricultural Production to Thrive Together.* Washington, DC: American Farmland Trust, 2020.

Spangler, Kaitlyn, Emily K. Burchfield, and Britta Schumacher. "Past and Current Dynamics of U.S. Agricultural Land Use and Policy." *Frontiers in Sustainable Food Systems* 4 (July 2020): 1–21.

Stoll, Steven. *Larding the Lean Earth: Soil and Society in Nineteenth-Century America.* New York: Hill and Wang, 2002.

Strange, Marty. *Family Farming: A New Economic Vision.* 2nd ed. Lincoln: University of Nebraska Press, 2008.

Successful Farming. "Farmer Suicides Today vs. 1980s Farm Crisis." April 30, 2018. https://www.agriculture.com/family/health-safety/sf-special-farmer-suicides-today-vs-1980s-farm-crisis.

Sue, Derald W., Christina M. Capodilupo, Gina C. Torino, Jennifer M. Bucceri, Aisha M. B. Holder, Kevin L. Nadal, and Marta Esquilin. "Racial Microaggressions in Everyday Life: Implications for Clinical Practice." *American Psychologist* 62, no. 4 (2007): 271–286.

Sumner, Daniel A. "American Farms Keep Growing: Size, Productivity, and Policy." *Journal of Economic Perspectives* 28, no. 1 (2014): 147–166.

Sutterfield, Ragan. *Wendell Berry and the Given Life.* Cincinnati, OH: Franciscan Media, 2017.

Swanson, Drew A. *A Golden Weed: Tobacco and Environment in the Piedmont South.* New Haven, CT: Yale University Press, 2014.

Szlanfucht, David L. "How to Save America's Depleting Supply of Farmland." *Drake Journal of Agricultural Law* 4, no. 1 (1999): 333–356.

Taylor, Dorceta E. "Black Farmers in the USA and Michigan: Longevity, Empowerment, and Food Sovereignty." *Journal of African American Studies* 22 (2018): 49–76.

———. *The Rise of the American Conservation Movement: Power, Privilege, and Environmental Protection.* Durham, NC: Duke University Press, 2016.

Thompson, Aaron W., and Linda Stalker Prokopy. "Tracking Urban Sprawl: Using Spatial Data to Inform Farmland Preservation Policy." *Land Use Policy* 26 (2009): 194–202.

Thompson, Charles D., Jr. *Going over Home: A Search for Rural Justice in an Unsettled Land.* White River Junction, VT: Chelsea Green, 2019.

Thompson, Paul B. *The Spirit of the Soil: Agriculture and Environmental Ethics.* 2nd ed. New York: Routledge, 2017.

Tolbert, Charles, Michael Irwin, Thomas Lyson, and Alfred Nucci. "Civic Community in Small-Town America: How Civic Welfare Is Influenced by Local Capitalism and Civic Engagement." *Rural Sociology* 67, no. 1 (2002): 90–113.

Tolnay, Stewart E. *The Bottom Rung: African American Family Life on Southern Farms.* Urbana: University of Illinois Press, 1999.

Touzeau, Leslie. " 'Being Stewards of Land Is Our Legacy': Exploring the Lived Experiences of Young Black Farmers." *Journal of Agriculture, Food Systems, and Community Development* 8, no. 4 (2019): 45–60.

TWRA (Tennessee Wildlife Resources Agency). *Tennessee State Wildlife Action Plan.* Nashville, TN: TWRA, 2015. https://www.tn.gov/twra/wildlife/action-plan/tennessee-wildlife-action-plan.html.

USDA (US Department of Agriculture). "Farm Household Well-Being: Glossary." Economic Research Service, May 24, 2022. https://www.ers. usda.gov/topics/farm-economy/farm-household-well-being/glossary/.

———. "Heirs' Property Landowners." Accessed February 17, 2022. https:// www.farmers.gov/working-with-us/heirs-property-eligibility.

———. "2017 Census of Agriculture Highlights—Black Producers." National Agricultural Statistics Service (NASS), 2019.

US Department of Energy, Office of Energy Efficiency and Renewable Energy. *Solar Futures Study*. Washington, DC: US Department of Energy, September 2021. https://www.energy.gov/eere/solar/solar-futures-study.

Usher, M. D., trans. *How to Be a Farmer: An Ancient Guide to Life on the Land*. Princeton, NJ: Princeton University Press, 2021.

Valliant, Julia C. D., Kathryn Z. Ruhf, Kevin D. Gibson, J. R. Brooks, and James R. Farmer. "Fostering Farm Transfers from Farm Owners to Unrelated, New Farmers: A Qualitative Assessment of Farm Link Services." *Land Use Policy* 86 (2019): 438–447.

van Wensveen, Louke. *Dirty Virtues: The Emergence of Ecological Virtue Ethics*. Amherst, NY: Humanity Books, 2000.

Verduyn, Phillipe, and Saskia Lavrijsen. "Which Emotions Last Longest and Why: The Role of Event Importance and Rumination." *Motivation and Emotion* 39, no. 1 (2015): 119–127.

Walker, Melissa. *Southern Farmers and Their Stories: Memory and Meaning in Oral History*. Lexington: University Press of Kentucky, 2006.

Wedell, Katie, Lucille Sherman, and Sky Chadde. "Midwest Farmers Face a Crisis: Hundreds Are Dying by Suicide." *USA Today*, March 9, 2020. https://www.usatoday.com/in-depth/news/investigations/2020/03/09/climate-tariffs-debt-and-isolation-drive-some-farmers-suicide/4955865002/.

Weible, Christopher M., and Paul A. Sabatier. "A Guide to the Advocacy Framework Coalition." In *Handbook of Public Policy Analysis: Theory, Politics, and Methods*, edited by Frank Fischer, Gerald J. Miller, and Mara S. Sidney, 123–136. Boca Raton, FL: CRC Press, 2007.

Weingarten, Debbie. "Why Are America's Farmers Killing Themselves?" *Guardian*, December 6, 2017. https://www.theguardian.com/

us-news/2017/dec/06/why-are-americas-farmers-killing-themselves-in-record-numbers.

Weissman, Sara. "A Debt Long Overdue." *Inside Higher Ed*, April 26, 2021. https://www.insidehighered.com/news/2021/04/26/tennessee-state-fights-chronic-underfunding.

White, Jeanne S. "Beating Plowshares into Townhomes: The Loss of Farmland and Strategies for Slowing Its Conversion to Nonagricultural Uses." *Environmental Law* 28, no. 1 (1998): 113–144.

White, Monica M. *Freedom Farmers: Agricultural Resistance and the Black Freedom Movement*. Chapel Hill: University of North Carolina Press, 2019.

Whitt, Christine E., Jessica E. Todd, and James M. MacDonald. "America's Diverse Family Farms, 2020 Edition." Economic Information Bulletin 220. US Department of Agriculture, Economic Research Service, December 2020.

Wickenden, Dorothy. "Wendell Berry's Advice for a Cataclysmic Age." *New Yorker*, February 28, 2022. https://www.newyorker.com/magazine/2022/02/28/wendell-berrys-advice-for-a-cataclysmic-age.

Wiebe, Joseph R. *The Place of Imagination: Wendell Berry and the Poetics of Community, Affection, and Identity*. Waco, TX: Baylor University Press, 2017.

Wirzba, Norman. "An Economy of Gratitude." In *Wendell Berry: Life and Work*, edited by Jason Peters, 142–155. Lexington: University Press of Kentucky, 2007.

Womack, Veronica L. "Black Power in the Alabama Black Belt to the 1970s." In *Beyond Forty Acres and a Mule: African American Landowning Families since Reconstruction*, edited by Debra A. Reid and Evan P. Bennett, 231–253. Gainesville: University Press of Florida, 2012.

Wood, Spencer D., and Jess Gilbert. "Returning African American Farmers to the Land: Recent Trends and a Policy Rationale." *Review of Black Political Economy* 27 (2000): 43–64.

Wozniacka, Gosia. "Is It a Farm If It Doesn't Sell Food?" *Civil Eats*, April 12, 2019. https://civileats.com/2019/04/12/ag-census-is-it-a-farm-if-it-doesnt-sell-food/.

Wright, D. Wynne. "Fields of Cultural Contradictions: Lessons from the Tobacco Patch." *Agriculture and Human Values* 22 (2005): 465–477.

Young, Mathias A. "Keep It in the Family: The Problems Facing Small Family Farms and How the Uniform Partition of Heirs' Property Act Can Help Owners Hold On to Generational Homesteads." *Current Issues Blog, Wake Forest Law Review*, November 11, 2021. http://www.wakeforestlawreview.com/2021/11/keep-it-in-the-family-the-problems-facing-small-family-farms-and-how-the-uniform-partition-of-heirs'-property-act-can-help-owners-hold-on-to-generational-homesteads/.

Young, Nicole. "Confusion over Greenbrier-Area Development Cleared by Robertson's Attorney." *Nashville Tennessean*, March 29, 2017. https://www.tennessean.com/story/news/local/robertson/2017/03/29/confusion-over-greenbrier-area-development-cleared-robertsons-attorney/99777274/.

Zagzebski, Linda Trinkaus. *Virtues of the Mind: An Inquiry into the Nature of Virtue and the Ethical Foundations of Knowledge*. Cambridge: Cambridge University Press, 1996.

Index

References to maps and figures are indicated by italicized page numbers.

virtues for good care and use of the
earth, 1–3, 19–22, 26, 36, 82, 103,
205–206n3, 219n5; terminology of
"environmental" vs. "ecological" and,
205n2; on University of Kentucky
faculty, 20; as writer, 20

Berry, Wendell, writings by: *The Art of
Loading Brush*, 207n21;
"Conservationist and Agrarian," 15,
90; *The Hidden Wound*, 166, 167,
220n10; "It All Turns on Affection,"
9; "Looking Ahead," 219n1;
"Manifesto: The Mad Farmer
Liberation Front" (poem), 2; "A
Native Hill," 214n4; "Notes from an
Absence and a Return," 88;
"Renewing Husbandry," 10; "A
Rescued Farm," 22–23; "Seven
Amish Farms," 25–26; "Stand by
Me," 219n5; "A Talent for Necessity,"
23–24; "Think Little," 29, 206n13;
The Unsettling of America, 11

Besuden, Henry, 23–24

Big Agriculture. *See* large-scale
agriculture

big row-crop farming, 23, 77, 145

Bilbro, Jeffrey, 10

Black Family Land Trust, 124, 179,
219–220n6

Black farmers: in 1920, 120; advocacy
for, 119–120, 198; affection, practice
of, 106–107, 147, 152–157, 167, 219n2;
civil rights advocacy and, 152, 177;
county agricultural committees,
underrepresentation on, 121;
disrespect of, 131–132; educational
and service extension programs for,
127–129, 185, 186; emancipation of
enslaved persons and, 117; endurance
through hardship, 158, 165, 167; as

exemplars for way forward, 168;
family legacy and, 106; farmland loss
and, 118, 122–124, 133–143, 167, 177;
farm size, 127, 178; federal and state
programs to support, 128, 177–179,
180–181; fidelity and, 106–107, 147,
157–162, 167; financial challenges for,
123–127, 145–146, 159, 162, 167;
"forty acres and a mule," 117, 216n13;
health problems of, 130; healthy race
relationships in rural communities,
132; historical background of, 107,
115–125; historically Black colleges
and universities' extension services
for, 127–128; as "historically
underserved," 173, 177, 197–198;
imagination and, 147, 148–152, 167;
liberation, farming representing, 165,
177; local measures to support,
185–186; in Maury County, 53,
106–107, 125–133, 147, 165, 167, 186;
microaggressions and, 130–131,
218n34; misrepresentation of,
131–132; mistrust of government
programs by, 128–129; outreach for
ethnographic studies, 108–109; pride
of, 144, 147, 150, 152, 162, 164; as
regenerative farmers, 116, 165, 177; in
Robertson County, 52–53, 106, 186;
sense of belonging and, 147; as
sharecroppers, 117, 119; spirituality
and, 106, 162–165; stewardship
virtues and motivations of, 107, 147,
162–163, 166, 171, 219n6; stress and
anxiety experienced by, 130; tax
payments by, 162–163; tenancy in
common and, 122–123;
underrepresentation among US
agricultural producers, 121–122, 165;
young Black people entering or

connection to the land: affection and,
92–93; grandparents and parents as
exemplars for, 83–84, 88, 150–151, 157,
160; interdependence between people
and the earth, 165–166, 215n14;
part-time farmers and, 90;
responsibility for well-being of the
land and, 103, 219n5. *See also*
motivations; multigenerational
connections to land; stewardship
virtues
conservation. *See* land protection/
conservation
conservation easements, xi–xii, 68, 85,
98–99, 171, 182, 214n11
Coopertown (Tennessee), 43, 65, 71
Cost of Community Services (CCS),
191–193, 222n27
cotton production, 116, 121
county agricultural committees/
commissions, 121
courage, 4, 197, 198
COVID-19 pandemic: agricultural
relief funds and, 178, 213n14;
economic challenges from, 94; effect
on ethnographic studies, 50–51, 108,
215n5; migration from urban areas
during, 140; school challenges from,
105
crop-insurance programs, 175, 180
Cross Plains (Tennessee), 43, 60,
65, 72

dairy farms, 44, 56, 84, 150, 210n2
Daniel, Pete: *Dispossession:
Discrimination against African
American Farmers in the Age of Civil
Rights*, 121
Daniels, Tom, 190, 212n4
Davidson County (Tennessee), 40–41

Davy, Dānia, 178
Denise (study participant), 57, 60,
71–72, 77, 96
despair and depression, 36, 71, 129–130,
143
Dillard, Annie, 205n3
Dioum, Baba, 205n3
discipline, 7–8, 12, 29
dispossession, 27, 123–124, 126, 167,
177, 180. *See also* farmland loss; heirs'
property
Dodson Gray, Elizabeth, 10
Du Bois, W. E. B., 118–119
Dudley, Kathryn: *Debt and
Dispossession: Farm Loss in America's
Heartland*, 27, 213n7, 213n9
Dyer, Janice, 123

easements. *See* conservation easements
economic challenges. *See* hardships,
endurance through; small and
midsized farmers
economic injustice, 59, 168, 172
economic motivation. *See* agricultural
viability; profit
education and training of farmers:
Black farmers' interest in, 150,
185–186; Black vs. white farmers,
127–128; financial education, 185,
186; mistrust of government
programs by Black farmers, 128–129
1890s schools, 127–128
Emancipation Proclamation, 117
Emerson, Ralph Waldo, xiii
eminent-domain exceptions, 188,
221–222n20
emotions: anger, 70, 130, 198; despair
and depression, 36, 71, 129–130, 143;
emotional toll of farmland loss,
70–75, *74*, 102, 109–113, *114–115*,

emotions (continued)
141–142, 190, 213nn7–8; fear about
future of the land, 159; positives of, 13
empathy, 12, 17, 28, 54, 132, 166, 168,
197–198
enduring hardships. *See* hardships,
endurance through; perseverance
engagement. *See* empathy
"enough" concept, 84
environmental benefits of farmland,
194–196
environmental ethics, 6–7; Berry and,
206n13; virtue-based, 206n13
environmental injustice, 169
environmentalism, 103–104, 194–195
environmental legislation, 1, 7, 206n13,
214n11
Environmental Quality Incentives
Program, 176
Equal Justice Initiative, 118
estate planning. *See* heirs' property;
succession planning
ethnographies: amateur, 21–26, 212n18;
COVID-19 pandemic's effect on
gathering, 50–51, 108; derision for
smaller farms, encounter with, 174;
formal ethnography, need for,
26–28, 67, 209n51; in Maury
County study, 108–109; methods
used in, 50–55, 215n5; in Robertson
County study, 49–55; shortcoming
of research in Robertson County,
52–53; snowball sampling used in,
52; storytelling as part of, 54, 70, 75.
See also qualitative research
exemplars, 22–26; of "a good farmer,"
29–32, 83–84, 97, 103, 207–208n22;
Aiken as, 22–23, 24; Berry as, 19–21;
Besuden as, 24; grandparents and
parents as, 9, 83–84, 88, 150–151, 157,

160; hooks as, 166; importance of,
30, 93, 150, 160, 164; Manier as, xiii;
strip-miner as model exploiter, 22
exploitation: curtailing, 171;
environmental, 168; by farmers, 2, 3,
30, 206n4; heirs' property structures
and, 123; by strip-miner, 22; of tax
breaks for farming, 46
extension agents and services, 23, 37,
108, 120, 127, 186
extraction: Berry arrested for protesting
mining practices, 29; protection of
natural resources, 172, 188;
strip-mining, 22; by the powerful, 1,
45–46, 59, 61, 118, 175, 181. *See also*
exploitation
"eyes to acres" ratio, 31

Fairbairn, Madeleine: *Fields of Gold*,
141
faith. *See* religious stewardship
motivations; spirituality
familial connections to land: Berry
family, 20; Black farmers and,
152–153, 162; hooks on, 218n39;
Manier family, xi–xiii; Yoder family,
25. *See also* multigenerational
connections to land; succession
planning
familiarity with place, xiii, 10–11, 20,
82, 150. *See also* attunement;
imagination; intimate knowledge
family farms. *See* small and midsized
farmers
farm auctions, 33–34, 35, 36, 67, 124,
155–156, 181, *183*
Farm Bill (2018), 171–172, 176, 220n3
Farm Bureau, 130
Farm Crisis (1980s), 27, 69, 73, 94,
213n9

Farmers Home Administration, 122
farmland consolidation. *See*
agricultural consolidation
farmland investing, 141
farmland loss: benefits lost due to,
193–194; of Black farmers, 118,
122–124, 133–143, 167, 177;
competition for land and, 140–141,
146; emotional reaction to, 70–75,
74, 102, 109–113, *114–115*, 141–142,
190, 213nn7–8; in Farm Crisis
(1980s), 27, 73, 213n9; federal actions
and policies to address, 171–180;
financialization of farmland and,
141; generational turnover and,
68–69, 123, 139, 157, 212n5; local
measures to address, 185–194; lot size
requirements as way to address, 187,
221n19; low-density residential
development and, 43–44, 64–66, *64*,
113–114, *113*, 138; in Maury County,
109–113, *114–115*, 133–143, *138*,
142–143; mini-farms carved out of
farmland, 46–48, *47*, 63–64, 113–114,
113, 137; partition sales and, 123–124;
permanence of, 72; prevalence of,
34–35, 57; rezoning and, 58–59; in
Robertson County, 37–44, 57–75,
74, 102, 109, 136; state actions and
policies to address, 180–185; state
rankings for, 210n1; unsolicited
offers for land, 57–58; urgency of
threat posed by, 169. *See also* farm
auctions; industrial and commercial
development; residential
development; urban sprawl
farm-link programs, 184–185
farm-service providers, 51–52, 59, 61,
108, 174
Farm Services Agency, 122

farm size: acreage statistics, 45, 112,
210n2; adding to family farm,
155–156; appropriate to locale, 31;
average farm size, 45; Black-operated
farms, 178; right-sized farmers, 32;
rise of big farms, 44–45, 48, 67, 75,
114, *115*; tiny-farm (mini-farm)
phenomenon, 46–48, *47*, 63–64,
113–114, *113*, 137
farm stores, 70, 108
farm transition. *See* succession
planning
Farrell, Justin, 213n7
federal actions and strategies for land
protection/conservation, 171–180,
184, 196; Agricultural Conservation
Easement Program (ACEP), 171–173;
bias toward big farmers, 173–176;
legislation proposed, 177–179
Federation of Southern Cooperatives,
124, 178
FFA. *See* Future Farmers of America
fidelity, 17–19; affection and, 16–17, 19,
21, 23, 28, 93, 99, 159; Aiken and, 23;
Berry and, 18, 20–21, 82, 94, 97, 160;
Besuden and, 24, 25; Black farmers
and, 106–107, 147, 157–162, 167;
continuing to practice and develop,
161, 169, 180, 199; definition of,
17–18, 157; essential to stewardship,
162; formal ethnography of, 26–28;
generational transfer of, 97–98,
160–162; home concept and, 158,
165, 219n4; hooks and, 166;
imagination and, 8, 82, 159; marriage
analogy to, 18, 21, 28, 94, 160–161; as
motivation for good care and use of
the earth, 160; power and potential
of, 170–171; as practical virtue, 3, 7,
17–19, 28–29; refusal to sell due to,

fidelity (continued)
95–97, 160–161, 164; of small and
midsized farmers, 36–37, 93–99;
Yoder family and, 26
Fifteenth Amendment, 117
fifty-year farm bill, 206n3
financing and mortgages for farms,
93–94, 167; bankruptcy and
repurchase of farm, 94–95; cloudy
title problems and, 181–182; federal
support, recommendations for,
171–180; financial education, 185,
186; racism in, 121, 123–127
Finney, Carolyn, 216n10
flood control, 195, 222n30
"forty acres and a mule," 117, 216n13
4-H, 185
Fourteenth Amendment, 117
Frances (study participant), 58, 61, 92,
184, 193
Francis, Charles, 191
Francis, Greg, 217n23
Frank (study participant), 57–58, 76, 96
Freedgood, Julia, 212n5
Freedmen's Bureau Act (1865), 117
Freedom Farm Cooperative, 125
Freilich, Robert, 211n5
friendship, 12–13, 17
Future Farmers of America (FFA), 89,
126, 151, 185, 186

Gallagher, Patrick, 195
Gates, Bill, 141
General Motors, 110, 133
generational turnover, 68–69, 123, 139,
157, 212n5. See also succession planning
generosity, 4–5
George (study participant), 131, 133,
137–141, 145, 148–149, 154–155,
159–160, 164–165

Georgia: farmland loss in, 34; freed
Black people receiving farmland in,
117; Uniform Partition of Heirs'
Property Act adopted in, 181
Gilbert, Jess, 120, 123–124
Giles County (Tennessee), 108
Gillibrand, Kirsten, 177
Global South, 207n20
good care of the earth. See motivations;
stewardship virtues
"goods," 4, 196; consumer goods,
production of, 209n54
Goodwin, Joy, 210n3
Gouldthorpe, Jessica, 210n3
Great Migration, 120
Great Recession, 34
Greenbrier (Tennessee), 65, 66
Green Revolution, 207n20
Greg (study participant), 70, 73–75, 83,
88
Grim, Valerie, 216n13

habitat loss and species endangerment,
195–196, 222n30
Hamer, Fannie Lou, 125
Hanson, J. D., 44
hardships, endurance through:
affection and, 14–15, 87–89, 92–93,
156; Black farmers and, 158, 165, 167;
fidelity and, 18–19, 94, 157, 160;
imagination and, 87, 151; moral
character acquired via, 6; small and
midsized farmers and, 114. See also
perseverance; physicality of farming
heirs' property, 122–125, 129–130, 156,
159, 177; Uniform Partition of Heirs'
Property Act, state adoption of,
180–181
Henry (study participant), 59–60, 100,
193

Maury County (Tennessee)
(continued)
ethnographic study of, 215n5;
"mini-farms" or large-lot residential
tracts, 113–114, *113*; motivations to sell
farmland, 139–140; open space,
benefits of, 194; residential
development, 111, *112*, 135–143, *136*,
138; similarities with Robertson
County, 109, 139–141, 144, 147, 149,
176; Walter's family farm, 105–106.
See also Black farmers; heirs' property
McFadden, Jonathan, 174–175
McGhee, Heather, 220n10
Melissa (study participant), 61–64, 68,
85, 98–99, 182, 214n11
Mennillo, Seth, 211n5
metis, 11
microaggressions, 130–131, 167, 218n34
Middle Tennessee State University
Center for Historic Preservation, 51–52
midsized farmers. *See* small and
midsized farmers
migration: Great Migration from the
South, 120; migrant farmworkers,
78; out-of-state migration to
Tennessee, 58, 134–135; urban
emigration during COVID-19
pandemic, 140; urban emigration to
become farmers, 31
millennials, 68
mini-farms, 46–48, *47*, 63–64, 113–114,
113, 137
Minnesota, farmers' connection to
land in, 27
Minorities in Agriculture and Natural
Resource Related Sciences
(MANRRS), 185
Mississippi, Uniform Partition of
Heirs' Property Act adopted in, 181

Mitchell, Thomas, 180–181
Montgomery County (Tennessee), 40
Moran, Greta: "Beginning Farmers,
Farmers of Color Outbid as
Farmland Prices Soar," 140
Moroney, Jillian, 27
Morrill Act: (1862), 216n13; (1890), 127,
216n13
mortgages. *See* financing and
mortgages for farms
motivations: for good care and use of
the earth, 1–2, 8, 14, 15, 26, 87,
90–92, 98, 103, 160, 219n5; resistance
to selling, 28, 95–97, 159–161, 164; to
sell farmland, 36, 67–68, 103, 139–140
Moton, Robert, 119
multigenerational connections to land,
25–26; affection and, 88; family
legacy and, 101, 106, 140, 165; fidelity
and, 97, 160; grandparents and
parents as exemplars, 83–84, 88,
150–151, 157, 160; imagination and,
83–84, 86; neighbors and family
members passing on knowledge, 9,
86, 97

Nashville: bedroom communities for,
43, 65–66, 110, 133; rezoning of
farmland, 59; urban sprawl, 34,
40–43, 60
Nathan (study participant), 68, 72,
85–86
National Endowment for the
Humanities. *See* Jefferson Lecture
Natural Resources Conservation
Service, 179
new farmers, 36, 44, 57, 70, 184
New York study on farmland's effect on
climate crisis, 194
New York University, 20

non-white farmers, 177. *See also* Black farmers; Indigenous people; racial diversity

North Carolina: ethnographic studies in, 215n5; farmland loss in, 34, 210n1; farmland protection in, 184, 221n15; Voluntary Agricultural Districts, 188

Northeast farm-link programs, 185, 221n16

off-farm employment and income, 36, 44, 58, 69, 73, 89, 93, 95, 100, 134, 140, 149, 164

Ohio: conservation easement programs in, 188; strip-mine conversion to farm in, 22

Olmstead, Grace, 2

Olson, Richard, 27

oneness with the land, 86–87, 149, 214n4. *See also* identity associated with place

open space, benefits of, 193–195. *See also* farmland loss

oral histories: transgenerational passage of, 86, 150, 153. *See also* storytelling

Orlinda (Tennessee), 43, 65

partition sales, 123–124. *See also* heirs' property; Uniform Partition of Heirs' Property Act

part-time farmers, 90, 125, 158. *See also* off-farm employment and income

Penniman, Leah, 125, 177

Pennsylvania, conservation easements in, 182

Perdue, Sonny, 2, 3

perseverance: affection's role in promoting, 15, 89–90; Black farmers and, 106, 147; fidelity and, 18, 99; in Robertson County, 37, 81, 89–90,

92, 101; stewardship virtues and, 169, 196

Pete (study participant), 72, 77, 84–85, 97–98

Petrusich, Amanda, 97, 207–208n22, 214n3

Phillip (study participant), 60–61, 83, 93, 95, 98

physicality of farming, 56, 69, 76, 89, 95, 139–140, 144, 148, 154, 158, 165

Pigford v. Glickman (1999, 2010), 122, 217n23

Pilgeram, Ryanne, 212n4, 213n7

place: affection for, 2, 16–17, 24; commitment to, 30, 169–170, 196; "culture of," 165–166, 168; familiarity with, xiii, 10–11, 20, 82, 150; identity associated with, 21, 86, 149, 164–165; sense of home and, 158, 165, 219n4; sense of imagination for, 9, 20

plantations. *See* slavery

Port Royal (Tennessee). *See* Spring Hill (Tennessee)

positivism, 207n20

poverty, 46, 122, 167

"Prairieview," 27

Presser, Lizzie, 125

Preston Jane, 211n8

pride of Black farmers, 144, 147, 150, 152, 162, 164, 166

profit: of farmers, 99–100, 144–146, 162; subordination of, 26. *See also* exploitation; motivations

protection of land. *See* land protection/ conservation

Prudential, 141; *See also* farmland investing

qualitative research, 49, 62, 65, 187, 213n7